T0211072

SpringerBriefs in Climate Studies

More information about this series at http://www.springer.com/series/11581

Daniel M. Alongi

Blue Carbon

Coastal Sequestration for Climate Change
Mitigation

 Springer

Daniel M. Alongi
Tropical Coastal & Mangrove Consultants
Annandale, QLD, Australia

ISSN 2213-784X   ISSN 2213-7858   (electronic)
SpringerBriefs in Climate Studies
ISBN 978-3-319-91697-2   ISBN 978-3-319-91698-9   (eBook)
https://doi.org/10.1007/978-3-319-91698-9

Library of Congress Control Number: 2018944674

Printed on acid-free paper

This Springer imprint is published by the registered company Springer International Publishing AG part of Springer Nature.
The registered company address is: Gewerbestrasse 11, 6330 Cham, Switzerland

Preface

The issue of blue carbon as a mitigation strategy for climate change impacts on global greenhouse gas emissions has been in place only for the past few years. Since 2009, there has been an explosion of scientific papers reporting on carbon stocks in a variety of estuarine and marine wetland habitats, especially tidal salt marshes, mangrove forests and seagrass meadows. Not surprisingly, there has been an explosion of websites, strategic ideas and pilot projects involving either restoration or replanting (or both) of these valuable coastal habitats. And while there have been a number of papers totalling up the amounts of carbon sequestered in soil and biomass in tidal wetlands, there is no complete overview of the entire issue or a critical look as to whether or not **REDD+** projects are truly worthwhile and certainly whether or not the money being put into such projects is money well spent.

The purpose of this book is to make a critical appraisal of this exploding ecological and climate change issue, a sort of 'stop and smell the roses' type of analysis and reflection on where the entire issue is headed. Indeed, the time is ripe for such a critical review as projects are being planned or run without a good understanding of the complexities of the issue of climate change adaptation and mitigation; there is a sense of rushing to judgement without a good sense of the intricacies and practicalities running such a project entails. As the reader will see, much has been learned by trial and error as the practical knowledge base expands and as naivety dissipates after the hard lessons have been learned.

Perhaps given the alluring nature of blue carbon as a panacea for climate change mitigation, it was inevitable that mistakes would be made in the early 'band wagon' days. But mistakes are still being made and some projects still lack good project planning and evaluation, and a good grounding in hard scientific reality. Not all lost habitat can be restored and some alternative sites have not been selected using the best or the most stringent set of scientific criteria.

Blue carbon is not a simple, linear course correction for losses or for partial destruction of habitat, as we all have much more to learn about how best to minimise mistakes and poor decisions, and how to maximise lessons learned the hard way and to best utilise the rapidly expanding state of knowledge. I hope that this slim volume will help in facilitating restoration and rehabilitation of these precious marine and coastal resources which continue to disappear at an alarming rate.

Acknowledgements

I thank Jim Fourqurean for help with the seagrass data, an anonymous reviewer for very helpful comments and my wife Fiona for the graphics.

Contents

Glossary

ACR	American Climate Registry
Anaerobic Soil	Soil which contains no free oxygen (aerobic soil)
Blue Carbon	Term coined to refer to the acquisition and storage of carbon in aquatic ecosystems especially in coastal habitats such as salt marshes, seagrass beds and mangrove forests
CAR	Carbon sequestration rate
CBD	Convention on Biological Diversity
CCB	The Climate, Community and Biodiversity Standard
CDM	Clean Development Mechanism
C_{org}	Organic carbon
FAO	Food and Agriculture Organisation of the UN
G	Gram
Gg	Gigagram $= 10^9$ grams
GCF	Green Climate Fund
GEF	Global Environment Fund
GHG	Greenhouse gases (carbon dioxide, methane, nitrous oxide)
GPA-Marine	Global Program of Action for the protection of the marine environment from land-based activities
IOC	International Oceanographic Commission
IPCC	Intergovernmental Panel on Climate Change
IUCN	International Union for Conservation of Nature and Natural Resources
LDCF	Least Developed Countries Fund
LULUCF	Land-Use, Land-Use Change and Forestry
MACs	Marginal abatement costs
MAR	Mass sediment accumulation rate
MEAs	Multilateral Environmental Agreement
Mg	Megagram $= 10^6$g or tonne
NAMAs	National Appropriate Mitigation Actions
NPV	Net present value

Oxidation	Loss of electrons such that CH_4 is oxidized to CO_2
PES	Payment for ecosystem services
Pg	Petagram $= 10^{15}$ g
PIC	Particulate inorganic carbon
POC	Particulate organic carbon
RAMSAR	Ramsar Convention on Wetlands
REDD+	Acronym for Reducing Emissions from Deforestation and Forest Degradation. The + refers to the additional steps of conservation and the sustainable management of forests and enhancement of carbon stocks
RSET	Rod surface elevation table, a method to estimate soil accretion rates in wetlands
SCC	Social cost of carbon
SCCF	Special Climate Change Fund
SDG	United Nation's Sustainable Development Goals
Tg	Teragram $= 10^{12}$ g
TNC	The Nature Conservancy
UNDP	United Nations Development Programme
UNEP	United Nations Environment Programme
UNESCO	United Nations Educational, Scientific and Cultural Organisation
UNFCCC	United Nations Framework Convention on Climate Change
VCS	Verified Carbon Standard

Chapter 1
Introduction

The term **blue carbon** was coined in November 2009 in a rapid response assessment report to a special inter-agency collaboration of the **UNEP**, **FAO** and **IOC/ UNESCO** (Nelleman et al. 2009). 'Blue carbon' is defined as the coastal carbon sequestered and stored by ocean ecosystems. The publication of the report was an important milestone as it completed the global carbon accounting assessment begun by the **IPCC** with atmosphere and terrestrial biomes.

The purpose of the report was to highlight the crucial role of the oceans and their ecosystems in maintaining earth's climate and to assist policymakers in focusing their discussions on adaption to and mitigating for climate change to the role of the oceans in emission reductions, as ocean ecosystems have been vastly overlooked.

The report had a number of recommendations for the protection, management and restoration of coastal ecosystems that are critical carbon sinks:

1. Establish a global blue carbon fund for protection and management of coastal and marine ecosystems and ocean carbon sequestration.
2. Immediately and urgently protect at least 80% of the remaining seagrass meadows, salt marshes and mangrove forests, through effective management.
3. Initiate management practices that reduce and remove threats and which support the robust recovery potential inherent in blue carbon sink communities.
4. Maintain food and livelihood security from the oceans by implementing comprehensive and integrated ecosystem approaches aiming to increase the resilience of human and natural systems to change.
5. Implement win-win mitigation strategies in the ocean-based sectors, including to improve energy efficiency in human-based uses (transportation, fishing, etc.) and to encourage sustainable ocean-based energy production; curtail unsustainable activities impacting on the oceans ability to absorb carbon; ensure investment for restoring and maintaining ocean carbon sinks; provide food and incomes that promote sustainable business development opportunities and catalyse the natural ability of coastal carbon sinks to regenerate by sustainable management practices.

© The Author(s) 2018
D. M. Alongi, *Blue Carbon*, SpringerBriefs in Climate Studies,
https://doi.org/10.1007/978-3-319-91698-9_1

Concurrently, a quantitative and qualitative assessment was commissioned by the **IUCN** (Laffoley and Grimsditch 2009) to document the carbon management potential of tidal salt marshes, mangrove forests, seagrass meadows, kelp forests and coral reefs. The report found that these habitats are quantitatively and qualitatively important, being highly valuable sources of food and fuel and for shoreline protection, and that all of them are amenable to management such as through marine protected areas, marine spatial planning and area-based fisheries management techniques. The key findings of the report were:

1. These key coastal ecosystems are of high importance because of the significant goods and services they provide as well as carbon management potential.
2. Their carbon management potential is equivalent to terrestrial ecosystems and may exceed the potential carbon sinks on land.
3. Coral reefs do not act as carbon sinks but are slight carbon sources due to their complex carbonate carbon chemistry.
4. In their analyses, salt marshes provide the greatest long-term rate of carbon accumulation in sediment ($210 \ gC \ m^{-2} \ year^{-1}$) compared with mangroves ($139 \ gC \ m^{-2} \ year^{-1}$) and seagrass meadows ($83 \ gC \ m^{-2} \ year^{-1}$). Data are insufficient to quantify the contribution of kelp forests.
5. The chemistry of sediments and soils from these ecosystems suggests that while small in geographical extent, the absolute comparative value of the carbon sequestered per unit area may be greater than similar processes on land, due to lower potential for emission of **GHG**s such as methane and carbon dioxide.
6. There is a lack of critical data for all habitats, especially those in tropical locations as having comprehensive carbon inventories is a critically important need.
7. These ecosystems are vital for food security of coastal inhabitants, especially in developing countries, for providing nursery grounds for artisanal fisheries and for also providing coastal protection by mitigating coastal erosion and storm surge so thus serving multiple functions in addition to carbon sequestration.
8. These habitats are endangered, with continuing losses and degradation, coupled with a lack of policy urgency to address current and future threats.
9. These ecosystems are threatened by nutrient and sediment run-off from land, displacement by urban development, aquaculture and overfishing, threatening their capacity to sequester carbon.
10. Management strategies need to be effective and strengthened by governments that already have commitments in place for biodiversity protection and sustainable development; such strategies however need to be enforced, especially in developing countries.
11. Anthropogenic GHG emissions are being underestimated because such emissions from these habitats are not being accounted for in national and international inventories, meaning that their carbon savings from sequestration do not count towards meeting national and international climate change commitments.

In 2010 a 'Blue Carbon Initiative' was established by the United Nations (via the IOC/UNESCO) in partnership with the Conservation International (**CI**) and the

International Union for the Conservation of Nature (**IUCN**). The aim of this initiative was to promote climate change mitigation through restoration and sustainable use of coastal and marine ecosystems. The initiative consists of two working groups, one on scientific and technical issues and the other on policy matters. The policy group (Herr et al. 2012) has made a number of recommendations to (1) integrate blue carbon activities fully into the international policy and financing processes of the United Nations Framework on Climate Change (**UNFCCC**) as part of mechanisms for climate change mitigation and into other carbon finance mechanisms such as the voluntary carbon market; (2) develop a network of blue carbon demonstration projects; (3) integrate blue carbon activities into other international, regional and national frameworks and policies; and (4) facilitate inclusion of the carbon value of coastal ecosystems in the accounting of ecosystem services.

In June 2012 at the Rio + 20 United Nations Conference on Environment and Development, the IOC released the *Blueprint for Ocean Sustainability* which contained two proposed measures to achieve ocean sustainability: the first relates to mitigating and adapting to ocean acidification, while the second proposes the creation of 'a global blue carbon market as a means of creating direct economic gain through habitat protection' (IOC 2011).

During this time, a report published in 2011 by the Nicholas Institute of Duke University in North Carolina, USA, came to similar conclusions to the two initial blue carbon reports that coastal habitats store large amounts of carbon in their soils and in biomass (Sifleet et al. 2011). The policy implications of blue carbon were the focus of this latter report which indicated that when these habitats are converted, their stored carbon is released back into the atmosphere as GHGs, thus reversing the effect of fostering carbon sequestration in **REDD+** and other rehabilitation projects. From a management point of view, salt marshes, mangroves and seagrass meadows should be considered in management of critical ecosystem services; one practical tool suggested was to pay landowners and managers for coastal blue carbon, assuming that protocols can be developed to allow these carbon stores to be traded on carbon markets.

The report reiterated that greater understanding of how such habitats sequester carbon, how to maximise such sequestration while minimising carbon losses, and where most of such sequestration is taking place, is urgently needed. Sifleet et al. (2011) also indicated that it is necessary to know how rapidly these habitats are being converted and the level of subsequent risk that carbon will be released back into the atmosphere from such activities, as well as the mechanisms and rate of CO_2 (carbon dioxide) and methane emissions that follow conversion of habitats. Policymakers need to understand that three components are involved in carbon sequestration and storage:

1. The annual sequestration rate: the yearly flux in a mature ecosystem of organic material transferred into **anaerobic soils** where it cannot undergo **oxidation** to CO_2 and be released into the atmosphere.
2. The amount of carbon stored in above- and below-ground biomass.
3. The total carbon stock stored in soils as a result of prior sequestration, that is, the historical sequestration over a particular habitat's lifetime.

The total carbon stock integrates the entire soil stock below-ground down to bedrock. This stock is a function of the soil carbon density and the soil depth. As Sifleet et al. (2011) pointed out, scientists have a better handle on density than the total depth as it is difficult to measure (by coring) soil profiles that in some habitats can be metres deep. While summarising available data, they concluded that the empirical database is poor and not representative of these habitats globally; information is biased in favour of some geographical regions than others. For example, salt marsh data is comparatively plentiful for North America and Europe but lacking for South America, Africa, Asia and other parts of the world. For mangroves, the situation is worse in that most data comes from Asia and Oceania, while for seagrasses, data is greatest for Europe and North America; there is a paucity of seagrass data from the tropics.

There is remarkably little data on the CO_2 emissions from habitats that have been converted. Most estimates have been made assuming that a certain percentage of biomass or soil lost is multiplied by their known carbon content. As McLeod et al. (2011) suggested, such assumptions and calculations may be highly inaccurate, and key questions need to be addressed:

1. How are sequestration rates and existing sediment carbon stocks affected by ecosystem loss and/or modification?
2. How may carbon sequestration rates and storage be affected by climate change?
3. What recommendations can be made to inform future carbon sequestration research?

The last question can be addressed as the need for better understanding of the drivers affecting carbon sequestration rates. For instance, possible drivers can be habitat age, temperature, primary productivity and respiration and their metabolic balance, soil or sediment type, carbon exchange with other ecosystems, location (estuarine vs marine), hydrology, sedimentation rate, differences in tidal elevation and in sea-level and species composition, to name but a few. Such information is essential for guiding the restoration and conservation of these habitats.

So, what is 'blue carbon? A conceptual model (Fig. 1.1) shows the major pathways of carbon flux among land, sea and atmosphere and indicates that most of the carbon buried in tidal salt marshes, mangrove forests and seagrass meadows is 'blue carbon'. The remainder is either emitted back to the atmosphere via respiration by plants and soil microbes and animals or exported to the ocean in the form of dissolved and particulate carbon. In the case of mangroves, substantial amounts of carbon may also be stored in above-ground biomass (but negligible in the other two habitats); thus the rates of gross and net primary productivity are an important feature (Chapter 3) because plant production leads to the increase over time in tree biomass. This presumes, of course, that the trees and timber above-ground are not destroyed or burned for human use.

Measuring and mapping spatial and temporal variations in carbon sequestration are also necessary in order to estimate and properly scale up as well as to relate these differences to physical, geological, chemical and ecological characteristics. It would also be useful to be able to assess and quantify land-use changes and to identify

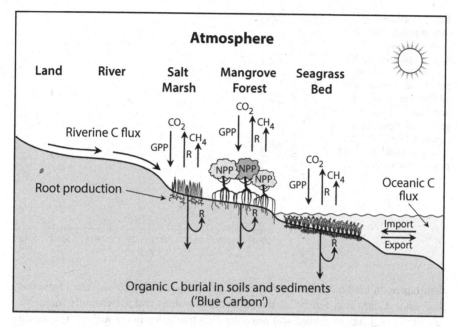

Fig. 1.1 Conceptual model of blue carbon in coastal ecosystems. *GPP* gross primary production, *NPP* net primary production, *R* respiration, CO_2 carbon dioxide, CH_4 methane

priority areas for conservation and management. The combined effects of climate change, land use and other human impacts on carbon sequestration need to be better understood.

Standardisation of methods is another important issue as the scientific literature is replete with papers in which different methods of soil and biomass sampling and chemical analysis have been used. Furthermore, it is important to verify rates of soil carbon burial as variable percentages of carbon deposited in soils are available to be oxidised to CO_2 mostly by complex and mixed microbial communities. These data are necessary to reveal the rate at which carbon is permanently buried in these habitats. Improved methods for measuring carbon storage can only help to inform regional and global carbon budgets.

Since the publication of these seminal papers and reports, there has been an explosion of subsequent papers on blue carbon (Fig. 1.2). This impressive growth was spawned by funding from various government and non-government agencies around the globe, as well as a lot of enthusiasm for the idea that the area as well as the quality of coastal wetlands could be expanded back to the time prior to the mid-twentieth century when destruction was not on an industrial scale (Hopkinson et al. 2012; Beaumont et al. 2014). Currently, these habitats are being lost at a rate of about 1% to 7% annually (Hopkinson et al. 2012). In fact, over the past 60–70 years, there has been a substantial increase in the amount of carbon and other materials transported from land to the ocean, indicating that both land and ocean habitats have seen a great deal of change. Regnier et al. (2013) estimated that anthropogenic

Fig. 1.2 The growth of blue carbon publications since 2009. The data are based on an exact phrase search for 'blue carbon' on Google Scholar and do not include all blue carbon-related publications as not all papers and reports (for instance, those of carbon stocks) used the keyword phrase 'blue carbon'

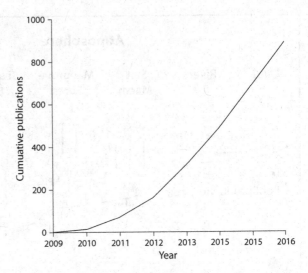

disturbance on land may have increased the flux of carbon to inland waters by about 20% since 1750. This can be attributed to deforestation and increasingly intensive cultivation that has increased soil erosion. This translates into roughly 40% being either emitted back to the atmosphere or buried in soils and sediment along the continuum from rivers to estuaries to coastal waters. As we shall see in the next four chapters, this increase has significant consequences for the amounts of blue carbon retained in the coastal zone.

References

Beaumont NJ, Jones L, Garbutt A, Hansom JD, Toberman M (2014) The value of carbon sequestration and storage in coastal habitats. Estuar Coastal Shelf Sci 137:32–40

Herr D, Pidgeon E, Laffoley D (eds) (2012) Blue carbon policy framework: based on the discussion of the International Blue Carbon Policy Working Group. IUCN, Gland/Arlington

Hopkinson CS, Cai WJ, Hu X (2012) Carbon sequestration in wetland dominated coastal systems – a global sink of rapidly diminishing magnitude. Curr Op Environ Sustain 4:186–194

IOC (2011) A blueprint for ocean and coastal sustainability. IOC/UNESCO, Paris

Laffoley D d'A, Grimsditch G (eds) (2009) The management of natural coastal carbon sinks. IUCN, Gland

McLeod E, Chmura GL, Bouillon S, Salm R, Björk M, Duarte CM, Lovelock CE, Schlesinger WH, Silliman BR (2011) A blueprint for blue carbon: toward an improved understanding of the role of vegetated coastal habitats in sequestering CO_2. Front Ecol Environ 9:552–560

Nelleman C, Corcoran E, Duarte CM, Valdés L, De Young C, Fonseca L, Grimsditch G (2009) Blue carbon: a rapid response assessment. United Nations Environment Programme, GRID-Arendal

Regnier P, Friedlingstein P, Ciais P, Mackenzie FT, Gruber N, Janssens IA, Laruelle GG, Lauerwald R, Luyssaert S, Andersson AJ, Arndt S, Arnosti C, Borges AV, Dale AW, Gallego-Sala A, Goddéris Y, Goossens N, Hartmann J, Heinze C, Ilyina T, Joos F, LaRowe DE, Leifeld J, Meysman FJR, Munhoven G, Raymond PA, Spahni R, Suntharalingam P, Thuller M (2013) Anthropogenic perturbation of the carbon fluxes from land to ocean. Nature Geosci 6:597–607

Sifleet S, Pendleton L, Murray BC (2011) State of the science on coastal blue carbon: a summary for policy makers. Nicolas Institute for Environmental Policy Solutions Report NI R 11-06. Nicolas Institute, Duke University, USA

Chapter 2
Salt Marshes

Salt marshes are intertidal wetlands that occur on low-energy shores and are comprised of herbaceous flowering plants and small scrubs. Being found mostly on sheltered coastal areas, salt marshes grow on silt and clay (mud) substrates. They occupy the interface between terrestrial and marine ecotones and have many attributes of both as well as some features that are unique (Adam 2016). The landward edge is delineated by sharp boundaries, but even in these cases, there is likely to be considerable freshwater groundwater flow linking marsh and land. There are often abrupt man-made boundaries, such as developments or even sand or rubble shingles where wrack material, such as marine detritus, often accumulates. The seaward boundary is usually delineated by sandflats, mudflats, seagrass beds and mangroves in subtropical and tropical regions.

Globally, salt marshes occur on all continents except Antarctica and are a general feature of temperate coastlines but can occur in low latitudes as a narrow fringe landward of mangroves and as more extensive stands on hypersaline flats where mangroves are excluded. These marshes are often sparse in terms of numbers of species present and biomass and may be interspersed with mats of microbes and microalgae. The global distribution of taxa shows that there are broad similarities between temperate marshes in the Northern and Southern Hemispheres and that major types of marsh can be related to latitude. Species richness of marsh grasses peaks not in the tropics, but in the high-latitude temperate zone.

Variations within an estuary can be large or subtle, depending on geomorphological, geological and chemical gradients. On average, changes up-estuary vary in relation to changes in salinity with more saline conditions at the mouth of the estuary to tidal freshwater marshes at the limit of saline intrusion. Brackish marshes predominate between both salt and freshwater endpoints, but these marshes are dominated by halophytic species with the saline zones being restricted to the higher less frequently inundated areas where evapotranspiration results in saltier conditions.

Salt marshes show zonation across the intertidal zone from low marsh to high marsh at the upper tidal limit. Davy (2000) has argued that zonation of salt marshes

© The Author(s) 2018
D. M. Alongi, *Blue Carbon*, SpringerBriefs in Climate Studies,
https://doi.org/10.1007/978-3-319-91698-9_2

is interpretive, being a spatial expression of succession of marsh species over time or distribution of species is the net result of environmental factors or competition over time. It is thus arguable as to whether or not zonation is an expression of succession or the result of environmental factors. Certainly, the boundaries between species or between communities are the net result of species' responses to environmental factors and to competitive interactions. Such patterns may change as a result of human interference in marsh hydrology, tectonic events, introduction of introduced species, or the result of climate change, but overall such patterns are stable over long stretches of time. Adam (2016) has suggested that zonation of marsh communities above the *Spartina* zone reflects succession from the 'original pre-*Spartina* pioneer' but that if a succession does occur as a result of progradation, it will be a new range of communities rather than a return to historic conditions.

2.1 Dynamics of Soil Accretion and Carbon Accumulation

The database on accretion rates in salt marshes is very large (unlike that for mangroves and seagrasses) and has been reviewed numerous times (Nixon 1980; Pethick 1981; Vernberg 1993; Allen 2000; Turner et al. 2001; Morris et al. 2002; Gedan et al. 2009; Shepard et al. 2011; Kirwan and Mudd 2012; Kirwan and Megonigal 2013). Accretion rates in salt marshes range from 2 to 10 mm year^{-1} with a median of 5 mm year^{-1}. Most marshes have rates of accretion of between 1 and 7 mm year^{-1}, a range which overlaps with that for mangrove forests (see Sect. 3.1.2).

The development of salt marshes reflects environmental history, but it is also a reflection of the net result of species' responses and interactions to the environment as well as to other species. Sea-level is the ultimate arbitrator of salt marsh delineation, and it is the considerable change in sea-level that has ultimately affected the spatial and long-term distribution of salt marsh environments. Sea-level rose to its present position about 6000 years ago following the last glacial maximum, and during that time, there was considerable change in the occurrence and extent of tidal marshes. As Adam (2016) points out, salt marshes can either erode away or develop quickly; as a marsh develops, it tends to prograde seawards and accretes. Both processes co-occur, but accretion can be maintained or even increased if sediment that is eroding or being transported downstream is deposited within the marsh.

Low- and mid-marsh zones can change rapidly due to changes in sedimentation as affected by either natural or anthropogenic processes, often within decades or one to two centuries, whereas upper marsh zones can remain stable for more than a millennium. Accumulation of sediment or soil results in an increase in surface height, and this accumulation may come about as a result of simple sediment accumulation or accumulation of organic material from plant roots or incorporated from burial of above-ground biomass, or both.

Like other ecosystems, marshes develop with the establishment of pioneer species which in turn promotes settlement of sediment particles. This positive feedback

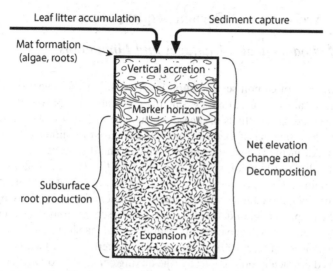

Fig. 2.1 The interrelationships between biotic and abiotic controls on soil accretion and elevation change. (Modified from Cahoon et al. 2006)

does not necessarily result in a uniform accretion of sediment, but sediment accumulation can be discontinuous in space and time. However, on average, further maturity of the marsh results in an increase in plant density and sediment retention, with a concomitant rise in surface elevation. This results in a negative feedback in which sedimentation begins to decline as the frequency of tidal flooding decreases. The net result is a rate of minerogenic (allochthonous) sedimentation of near zero at the tidal limit. Allochthonous sedimentation operates in two ways: (1) through capture of fine silt and clay particles by the leaves and stems and (2) settling out by the slowing of water motion through the marsh (Weis 2016). However, there are many ways in which tidal elevation can increase or decrease in salt marshes in addition to sediment inputs.

Figure 2.1 illustrates the key biotic and abiotic processes controlling soil accretion in salt marshes and in mangroves and seagrasses. Accretion processes in salt marshes are complex, but overall, the deposition of suspended particles during flooding, the accumulation of plant material (roots and litter) and the formation of algal mats and subsurface roots are the key inputs. Sediment capture is an abiotic process, although the slowing of tidal water flow by the presence of plants can be looked at as an indirect biotic process as well (Fig. 2.1). Decomposition by microbes and benthic infauna and compaction negatively influence net elevation change. The rate and quality of marsh accretion are influenced by both positive (ice rafting, flooding by storms, vegetative growth, CO_2 increase) and negative (auto-compaction, ice shearing, storm erosion) forces that on average have resulted in salt marsh accretion keeping pace with current rates of sea-level rise in most but not all regions (Weis 2016).

2.2   Soil Accretion and Carbon Sequestration

2.2.1   Methodological Advantages and Limitations

The measurement of carbon sequestration involves the measurement of soil accretion, and rates are as varied among salt marshes (and mangroves and seagrass meadows) as there are methods of measurement. The rate of soil accumulation can be measured by a number of methods, including the use of radiotracers and numerical models. All methods have inherent advantages and disadvantages.

A commonly used method is the use of ^{210}Pb and ^{137}Cs radioisotope geochronology (Robbins 1978). The method involves the measurement of naturally occurring and artificial radionuclides with increasing sediment depth and deriving a slope (rate) from a simple model of decrease in radioisotope concentration with increasing depth. A sediment core is taken carefully within a given marsh area and subsectioned usually to a depth of at least 100 cm. This depth is preferred as surface sediments are usually mixed physically or biologically (the mixing depth) destroying the pattern of probable decrease in concentration. Mass accumulation rates are derived from modelling the decline below the mixing depth of the radionuclides ^{210}Pb and ^{137}Cs, the latter derived from radioactive atomic bomb fallout. ^{210}Pb occurs naturally from cosmic radiation of the earth, and what is measured is both the naturally occurring isotope and that derived from the ^{226}Ra parent (the difference is called 'excess' ^{210}Pb). The three isotopes ^{226}Ra, ^{210}Pb and ^{137}Cs are measured by gamma spectroscopy in the sequentially sliced sediment samples. Both ^{210}Pb and ^{137}Cs profiles provide a check on the other to derive a mass sediment accumulation rate or **MAR**. This is done by interpreting the radiochemical tracer profiles for sedimentation history using models, such as those by Robbins (1978), which utilise a sediment mixed layer thickness, a decadal-century scale average input of excess ^{210}Pb (total ^{210}Pb minus parent ^{226}Ra) and diffusion coefficients for ^{210}Pb and ^{137}Cs in marine sediments. The models all utilise a regression of concentration by sediment depth and identify the mixed layer in the upper profile that does not fit the regression. The values obtained are then multiplied by bulk density and carbon content to derive a carbon sequestration rate.

The advantages of the method are that it does appear to meet the conditions of 'steady state' between inputs from particle scavenging and outputs from mixing and provides for an accurate estimate of mass sedimentation on the scale of decades to as much as a century. The other advantage is that all of this information can be gleaned from a single sediment core taken in a matter of minutes and with a minimum of field equipment. The disadvantages are that the gamma counting is time-consuming and the gamma equipment is expensive. There is also the problem that some cores, especially those taken in biologically and geologically active wetlands, are completely mixed, making interpretation of the radiochemical data impossible. Other disadvantages are that the surface sediments may have been disturbed or eroded by storms or runoff, compaction distorts the interpretation of the profiles and conditions of 'steady state' of radionuclide fluxes in fluvial environments are often not met. The interpretation of these profiles, even when clear ones have been obtained, is that the method relies on a number of assumptions such as the rate of scavenging of ^{210}Pb and ^{137}Cs onto suspended sediment particles versus dissolution in seawater is in chemical equilibrium. Further, the profiles can vary among replicate cores owing to small-scale differences in sedimentary history and grain size.

The radiochemical profiles must be considered cautiously as one must consider the scale limitations and the role of geologic history. For instance, one may obtain a MAR from a given location in an estuary, but these sediments may originate from another location within the same estuary, giving a false picture of overall net sediment accumulation when in fact the estuary sediment budget is in balance or may even showing a net loss of sediment to the sea.

Time scales longer than a century may be needed to properly assess MAR in estuaries with complex histories. Differences in tides may also affect mass sediment accumulation as well as relative changes in sea-level. We know that there was relatively slow sea-level rise in the late Holocene (1 mm year^{-1}) to rapid sea-level rise beginning in the eighteenth or nineteenth century. Over the past century, sea-level rise has averaged 2 mm year^{-1}.

Another common method to measure mass sediment accretion may help as it incorporates the entire sedimentary history of a given location. This is the rod surface-elevation table or **RSET** method (Cahoon et al. 2002a, b). The only disadvantage is that it is initially labour-intensive. The major advantage is that it is relatively inexpensive, portable and easy to measure. The rod surface-elevation table is a balanced, lightweight mechanical levelling device that attaches to both shallow (<1 m) and deep (preferably driven to bedrock) rod bench marks and is an advance on the original surface elevation table (SET) in allowing determination of elevation change occurring over different depths of the sediment profile. This improvement was made since it was discovered that subsurface processes such as root growth can exert a significant change over sediment elevation. For instance, one can measure the growth on elevation of the subsurface root zone versus the entire sediment profile. The rod provides confidence intervals for the height of an individual measurement in the laboratory of ±1.0 and 1.5 mm of two different operators. In the field, the confidence intervals are greater but within a few mm, making it very precise. And carbon sequestration rate can be derived by multiplying the sedimentation rate by bulk density and carbon content.

A third, but less common, method to determine sediment accretion is the use of numerical modelling of empirical measurements of suspended sediment concentrations within an estuary (Morris et al. 2002; Temmerman et al. 2003). For instance, in the Temmerman method, the rate of marsh accretion is assumed to be in steady state:

$$dE/dt = dS_{MIN}/dt + dS_{ORG}/dt - dP/dt \qquad (2.1)$$

where the first term is the rate of mineral sediment deposition (S_{MIN}) plus the second term which is the rate of organic accretion (S_{ORG}) minus the rate of compaction of the deposited sediment (P). In practical terms, the size of the second and third terms is negligibly relative to the mineral sediment component.

The first term dS_{MIN}/dt is derived from the settling velocity and the concentration of suspended mineral sediment particles:

$$dS_{MIN}/dt = \int\limits_{Year} \int\limits_{T} W_s \, x \, C(t)dt/p \qquad (2.2)$$

where W_s is the settling velocity, C is the depth-averaged suspended sediment concentration and p is the dry bulk density. The right-hand terms are integrated over the total duration (t) of a tidal inundation cycle and subsequently over all inundation cycles during a year. Temporal variation in suspended sediment concentrations (C_t) during a tidal inundation cycle is calculated by solving the following mass balance equation over a tidal cycle:

$$d[h(t) - E]C_t/dt = -W_sC(t)fC(o)dh/dt \qquad (2.3)$$

where h and f are the water surface and marsh surface elevations and $C(o)$ is the incoming suspended sediment concentrations in the water that floods the marsh surface. During flood tide, when $dh/dt > 0$, the incoming suspended sediment concentration ($C(o)$) is proportional to the inundation depth at high tide [$h (t_{HW}) - E$]:

$$C(o) = k[h(t_{HW}) - E] \qquad (2.4)$$

while during ebb tide (when $dh/dt < 0$), $C(o)$ is set to equal $C(t)$ and k is the proportionality constant calculated based on average suspended sediment concentrations over a long (weeks to months) period encompassing spring and neap tides.

The advantage of this method is that one can hindcast and forecast sediment accretion with changes in predicted sea-level rise. The disadvantage is that one needs the empirical suspended sediment data to run the model and some basic knowledge of modelling and mathematics.

2.2.2   Carbon Sequestration Rates

The data on carbon sequestration rates (**CAR**) in salt marshes (Fig. 2.2) shows no clear relationship with changes in latitude as it is likely that these rates are a function of a number of interrelated factors, such as marsh age, tidal inundation frequency, tidal elevation, marsh geomorphology, species composition, soil grain size, catchment and river input, ocean input and degree of human impact. The mean ($\pm$ 1SE) of 212 ± 18 **g C$_{org}$** m^{-2} year^{-1} ($n = 168$ locations; median = 184; minimum = 9; maximum = 1713) represents a wide array of salt marsh locations, types and ages. Ouyang and Lee (2014) indicate that latitude, tidal range and elevation appear to be the most important drivers of CAR, with considerable variation among biogeographic provinces. Species composition may play an important role as Ouyang and

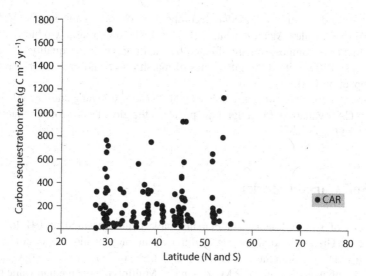

Fig. 2.2 Rates of carbon sequestration in salt marshes as a function of latitude. (Sources: Data and references in Chmura et al. (2003); Lovelock et al. (2014); Ouyang and Lee (2014); Weston et al. (2014); Yuan et al. (2014); Artigas et al. (2015); Davis et al. (2015); Ye et al. (2015); Kelleway et al. (2016); Wang et al. (2016); Macreadie et al. (2017a, b))

Lee (2014) calculated that *Spartina* marshes have significantly higher CAR but no significant differences among marshes composed of *Phragmites, Juncus, Halimione* and *Distichlis*. There is some evidence that European marshes support higher CAR, although there is a lack of significant differences among European, tropical West Atlantic, Northwest Atlantic and Northeast Pacific marshes; too few data come from Australasia, China, Japan and the Arctic.

The latitudinal pattern for salt marshes (Fig. 2.2) is unclear, but the few data from north of latitude 68.4° suggests lower CAR (Ouyang and Lee 2014), while those within the band of 48.4–58.4°N show the highest but very variable rates of CAR. This can be interpreted as reflecting a slower and shorter growing season at boreal latitudes as well as a difference in the balance between rainfall and evaporation.

It is most likely that tidal range and tidal elevation of a given marsh have more to do with driving CAR than any other factors based on the simple premise that more frequent inundation with sediment-laden tidal waters leads to more sediment available to deposit. Ouyang and Lee (2014) calculated that tidal range accounts for nearly 52% of the variation in CAR, as opposed to latitude which accounts for only 29.6% of variation. Soil CAR shows a clear decline from low to high marsh sites reflecting the more frequent inundation and subsequent settling of sediment on the marsh surface. Root production is higher in high versus low marsh locations, but this may be offset by the lower sediment accretion rate and bulk density of soil towards land.

Of course, other factors drive CAR such as below-ground root production. Soil CAR for salt marshes has been shown to be positively related to below-ground productivity and negatively related to decomposition of soil organic matter (McLeod et al. 2011; Gonzalez-Alcarez et al. 2012). Both below-ground productivity and organic matter decomposition are affected by tidal regime. Litter quality and quantity may also differ with different species of marsh grass, and this factor may have some impact on CAR.

Assuming a total area of salt marshes of 41,657 km^2 (Ouyang and Lee 2014) and weighing CAR values by biogeographic province the global CAR for salt marshes is about 10 Tg C$_{org}$ year^{-1}.

2.3   Soil Carbon Stocks

The stocks of soil carbon in salt marshes have been measured at 191 locations (Table 2.1). There are no significant differences among locations (excluding locations of $n = 1$). Based on these data, a mean ($\pm$ 1 SE) of 317.2 $\pm$ 19.1 **Mg** C$_{org}$ ha^{-1} is derived with a median of 282.2 Mg C$_{org}$ ha^{-1}. Multiplying the median value by the global area of salt marshes (41,657 km^2; Ouyang and Lee 2014) gives a total global estimate of 1.2 **Pg** C$_{org}$ (=1156 **Tg** C$_{org}$) of salt marsh carbon. Comparatively few papers provide both above- and below-ground carbon stocks, but, on average, 99.2% of total carbon is vested below-ground in soils and to a much lesser extent in roots (Kennedy et al. 2014). Thus, virtually all ecosystem carbon is below-ground in salt marshes.

As gleaned from the range of values for each location, there is no clear or simple relationship of carbon stocks with any particular environmental or ecological factor except for tidal elevation and frequency of tidal inundation. Overall, the same discussion for CAR applies for carbon stocks as there is a trend of greater carbon stocks in low versus high marshes. Thus it appears that higher CAR results in greater carbon stocks.

Carbon storage is highest in mature salt marshes that have been stable or pristine, or both, for a long time, usually centuries (Artigas et al. 2015). Lower carbon stocks are found in newly restored marshes, such as those in river deltas in China (Yu et al. 2013; Zheng et al. 2013; Yuan et al. 2014; Ye et al. 2015). In a number of countries, marshes have been reclaimed and thus drained and diked to prevent tidal flooding to create good conditions for land agriculture (Connor et al. 2001), and it is in these marshes, even when restored, where carbon stocks have been deleted by disruption of the upper soil layers where most of the organic carbon occurs. Low stocks also occur in marshes where erosion is common (Wang et al., 2016), where mangrove encroachment is occurring (Kelleway et al. 2016) and in marsh soils that are acidic due to high rates of organic matter decomposition (Ye et al. 2015).

Globally, wetlands have lost significant amounts of organic carbon historically, about 55 Pg for the world's soils, including those tilled for agriculture. For instance, China has lost large swaths of freshwater and estuarine marsh, about 70 Tg since the 1970s (Zheng et al. 2011). This situation is similar for Japan and other Asian countries that converted wetlands for dry land agriculture.

Table 2.1 Estimates of organic carbon stocks (Mg C_{org} ha^{-1}) in salt marsh soils to a depth of 1 m

Location [Sources]	Number of observations	Range	Mean
Northeast Canada[a]	38	170–735	343
China[b]	28	27–1560	189
Gulf of Mexico[c]	27	31–1900	500
New England[d]	23	120–600	356
United Kingdom[e]	17	64–576	225
Chesapeake Bay[f]	13	188–540	330
California[g]	11	79–433	234
Australia[h]	8	61–343	171
North Carolina[i]	6	47–556	316
Florida[j]	5	155–463	278
Netherlands[k]	4	200–410	327
Georgia[l]	4	200–233	213
United Arab Emirates[m]	3	30–164	80
Denmark[n]	2	210–270	240
Delaware Bay[o]	2	143–253	198
France[p]	1		730
British Columbia[q]	1		170
Spain[r]	1		30
New Jersey[s]	1		1373

Updated from Chmura et al. (2003) and Sifleet et al. (2011)

[a]Sources: Chmura et al. (2003) and Chmura and Hung (2004)

[b]Sources: Yu et al. (2013), Zheng et al. (2013), Yuan et al. (2014), Ye et al. (2015) and Wang et al. (2016)

[c]Sources: Hatton et al. (1983), Callaway et al. (1997), Bryant and Chabreck (1998), Markewich (1998), Chmura et al. (2003) and Campos et al. (2011)

[d]Sources: Howes et al. (1985), Roman et al. (1997), Orson et al. (1998), Anisfield et al. (1999), Chmura et al. (2003) and Johnson et al. (2007)

[e]Sources: French and Spencer (1993), Callaway et al. (1997), Chmura et al. (2003) and Cantarello et al. (2011)

[f]Sources: Kearney and Stevenson (1991), Blum (1993), Chmura et al. (2003), Hussein et al. (2004) and Thomas and Blum (2010)

[g]Sources: Cahoon et al. (1996), Chmura et al. (2003), Brevik and Homburg (2004), Drexler (2011) and Callaway et al. (2012)

[h]Sources: Livesley and Andrusiak (2012), Saintilan et al. (2013), Kelleway et al. (2016) and Macreadie et al. (2017b)

[i]Sources: Craft et al. (1988, 1993) and Chmura et al. (2003)

[j]Sources: Choi and Wang (2004) and Chmura et al. (2003)

[k]Sources: Buth (1987), Callaway et al. (1996) and Chmura et al. (2003)

[l]Sources: Loomis and Craft (2010) and Więski et al. (2010)

[m]Source: Schile et al. (2017)

[n]Sources: Morris and Jensen (1998) and Chmura et al. (2003)

[o]Source: Weston et al. (2014)

[p]Sources: Hensel et al. (1999) and Chmura et al. (2003)

[q]Source: Chmura et al. (2003)

[r]Source: Curado et al. (2013)

[s]Source: Artigas et al. (2015)

Indirect human activities can also negatively impact marsh accretion. In a study of impacted marshes on Cape Cod, Massachusetts, Coverdale et al. (2014) found that, using historical records and field experiments, salt marshes have lost significant amounts of soil and carbon over two centuries as a result of direct and indirect human impacts. Direct impacts include habitat conversion, boating and dredging, while indirect impacts include eutrophication, die-off, oiling and mosquito ditching. Some losses do occur naturally, such as erosion and calving, facilitated by crab burrows and marsh die-off. Over three decades, they found that creek banks in die-off marshes retreated more rapidly than healthy creek banks. Indirect impacts on salt marshes greatly alter ecological processes that tend to sequester soil and carbon, shifting from a natural bottom-up to top-down control which increases marsh vulnerability to die-off.

2.4 Potential Losses of Carbon

The loss of carbon shifts marshes from net sinks to net sources of carbon (as carbon dioxide and methane) for the atmosphere. McLeod et al. (2011) estimates that since the 1800s, 25% of global salt marsh area has been lost. Assuming that the global median carbon stock is 282 Mg C_{org} ha^{-1}, that 95% of all carbon has been oxidised to CO_2 (Kennedy et al. 2014) and that 1,150,000 ha of salt marsh has been destroyed, these losses equate to 1.1 Pg CO_2e (equivalents) returned to the atmosphere or coastal ocean since the 1800s. This figure must be considered minimal because soil deeper than 1 metre is usually lost on conversion and degradation.

Pendleton et al. (2012) derived a wider suite of numbers for current loss rates by assuming a 1–2% annual conversion/loss rate of salt marshes. They estimate that from 237 to 949 Mg CO_2e ha^{-1} is susceptible, with potential carbon emissions from all marsh losses and degradation from 20 to 240 Tg CO_2e year^{-1}. These values are highly variable owing not only to the range of 1–2% conversion, but also assuming a wide global extent of salt marshes from 2,200,000 to 4,000,000 ha^{-1}. Assuming that all salt marshes were destroyed and assuming the above-mentioned median C stock value and the wide range of salt marsh area remaining, the loss of marsh can result in 2.1 to 38.3 Pg CO_2 equivalents lost to the atmosphere which is equivalent to 0.5 to 8.3 years of emissions from global forest loss (Hansen et al. 2013). These values are only crude estimates, but all point to the fact that the potential return of carbon to the atmosphere upon loss or degradation of marsh habitat is potentially huge.

References

Adam P (2016) Saltmarshes. In: Kennish MJ (ed) Encyclopedia of estuaries. Springer, New York, pp 515–535

Allen JRL (2000) Morphodynamics of Holocene salt marshes: a review from the Atlantic and North Sea coasts of Europe. Quart Sci Rev 19:1155–1231

Anisfield SC, Toben MJ, Benoit G (1999) Sedimentation rates in flow restricted and restored salt marshes in Long Island Sound. Estuaries 22:231–244

Artigas F, Shin JY, Hobble C, Marti-Donati A, Schäfer KVR, Pechmann I (2015) Long term carbon storage potential and CO_2 sink strength of a restored salt marsh in New Jersey. Agricult Forest Meteorol 200:313–321

Blum LK (1993) *Spartina alterniflora* root dynamics in a Virginia marsh. Mar Ecol Prog Ser 102:677–718

Brevik EC, Homburg JA (2004) A 5000 year record of carbon sequestration from a coastal lagoon and wetland complex, Southern California, USA. Catena 57:221–232

Bryant JC, Chabreck RH (1998) Effects of impoundment on vertical accretion of coastal marsh. Estuaries 21:416–422

Buth GJC (1987) Decomposition of roots of three plant communities in a Dutch salt marsh. Aq Bot 29:123–138

Cahoon DR, Lynch JC, Powell AN (1996) Marsh vertical accretion in a southern California estuary, USA. Estuar Coast Shelf Sci 43:19–32

Cahoon DR, Lynch JC, Hensel P, Perez BC, Boumans RM, Day JW Jr (2002a) High-precision measurements of wetland sediment elevation: I. Recent improvements to the sedimentation-erosion table. J Sed Res 72:30–33

Cahoon DR, Lynch JC, Perez BC, Segura B, Holland RD, Stelly C, Stephenson G, Hensel P (2002b) High-precision measurements of wetland sediment elevation: II. The rod surface elevation table. J Sed Res 72:734–739

Cahoon DR, Hensel PF, Spencer T, Reed DJ, McKee KL, Sainilan N (2006) Coastal wetland vulnerability to relative sea-level rise: wetland elevation trends and process controls. In: Verhoeven JTA, Beltman B, Bobbink R, Whigham DF (eds) Wetlands and natural resource management. Springer, Berlin, pp 271–292

Callaway JC, Nyman JA, DeLaune RD (1996) Sediment accretion in coastal wetlands: a review and simulation model of processes. Curr Top Wetl Biogeochem 2:2–23

Callaway JC, DeLaune RD, Patrick WH Jr (1997) Sediment accretion rates from four coastal wetlands along the Gulf of Mexico. J Coast Res 13:181–191

Callaway JC, Borgnis EL, Turner RE, Milan CS (2012) Carbon sequestration and sediment accretion in San Francisco Bay tidal wetlands. Estuar Coast 35:1163–1181

Campos CA, Hernández ME, Moreno-Casasola P, Espinosa EC, Robledo RA, Infante Mata D (2011) Soil water retention and carbon pools in tropical forested wetlands and marshes of the Gulf of Mexico. Hydrolog Sci J 56:1388–1406

Cantarello E, Newton AC, Hill RA (2011) Potential effects of future land-use change on regional carbon stocks in the UK. Environ Sci Pol 14:40–52

Chmura GL, Hung GA (2004) Controls on salt marsh accretion: a test in salt marshes of Eastern Canada. Estuaries 27:70–81

Chmura GL, Anisfeld SC, Cahoon DR, Lynch JC (2003) Global carbon sequestration in tidal, saline wetland soils. Glob Biogeochem Cycles 17:1111

Choi Y, Wang Y (2004) Dynamics of carbon sequestration in a coastal wetland using radiocarbon measurements. Global Biogeochem Cycles 18. https://doi.org/10.1029/2004GB00224

Connor RF, Chmura GL, Beecher CB (2001) Carbon accumulation in Bay of Fundy salt marshes: implications for restoration of reclaimed marshes. Global Biogeochem Cycles 15:943–954

Coverdale TC, Brisson CP, Young EW, Yin SF, Donnelly JP, Bertness MD (2014) Indirect human impacts reverse centuries of carbon sequestration and salt marsh accretion. PLoS ONE 9:e93296

Craft CB, Broome SW, Seneca ED, Showers WJ (1988) Estimating sources of soil organic matter in natural and transplanted estuarine marshes using isotopes of carbon and nitrogen. Estuar Coast Shelf Sci 26:633–641

Craft CB, Seneca ED, Broome SW (1993) Vertical accretion in microtidal regularly and irregularly flooded estuarine marshes. Estuar Coast Shelf Sci 31:371–386

Curado G, Rubio-Casal AE, Figueroa E, Grewell BJ, Castillo JM (2013) Native plant restoration combats environmental change: development of carbon and nitrogen sequestration capacity using small cordgrass in European salt marshes. Environ Monit Assess 185:8439–8449

Davis JL, Currin CA, O'Brien C, Raffenburg C, Davis A (2015) Living shorelines: coastal resilience with a blue carbon benefit. PLoS ONE 10:e0142595. https://doi.org/10.1371/jour nal.pone.0142595

Davy AJ (2000) Development and structure of salt marshes: community patterns in time and space. In: Weinstein MB, Kreeger DA (eds) Concepts and controversies in tidal marsh ecology. Kluwer, Dordrecht, pp 137–156

Drexler JZ (2011) Peat formation processes through the millennia in tidal marshes of the Sacramento-San Joaquin delta, California, USA. Estuar Coasts 34:900–911

French JR, Spencer TS (1993) Dynamics of sedimentation in a tide-dominated back barrier salt marsh, Norfolk, UK. Mar Geol 110:315–331

Gedan KB, Silliman BR, Bertness MD (2009) Centuries of human-driven change in salt marsh ecosystems. Annu Rev Mar Sci 1:117–141

González-Alcaraz EC, Jiménez-Cárceles FJ, Párraga I, Maria-Cervantes A, Delgado MJ, Álvarez-Rogel J (2012) Storage of organic carbon, nitrogen and phosphorus in the soil-plant system of *Phragmites australis* stands from a eutrophicated Mediterranean salt marsh. Geoderma 185–186:61–72

Hansen MC, Potapov PV, Moore R, Hancher M, Turubanova TA, Thau D, Stehman SV, Goetz SJ, Loveland TR, Kommareddy A, Egorov A, Chini L, Justice CO, Townshend JRG (2013) High-resolution global maps of 21st-century forest cover change. Science 342:850–853

Hatton RS, DeLaune RD, Patrick W (1983) Sedimentation, accretion and subsidence in marshes of Barataria Basin, Louisiana. Limnol Oceanogr 28:499–502

Hensel PE, Day JW, Pont D (1999) Wetland vertical accretion and soil elevation change in the Rhone River delta, France: the importance of riverine flooding. J Coast Res 3:668–681

Howes BL, Dacey JWH, Teal JM (1985) Annual carbon mineralization and belowground production of *Spartina alterniflora* in a New England salt marsh. Ecology 66:595–605

Hussein AH, Rabenhorst MC, Tucker ML (2004) Modeling of carbon sequestration in coastal marsh soils. Soil Sci Soc Am 68:1786–1795

Johnson BJ, Moore KA, Lehmann C, Bohlen C, Brown TA (2007) Middle to Late Holocene fluctuations of C3 and C4 vegetation in a northern New England salt marsh, Sprague Marsh, Phippsburg, Maine. Org Geochem 38:394–403

Kearney MS, Stevenson JC (1991) Island land loss and marsh vertical accretion rate evidence for historical sea-level changes in Chesapeake Bay. J Coast Res 2:403–415

Kelleway JJ, Saintilan N, Macreadie PI, Skilbeck CG, Zawadzki A, Ralph PJ (2016) Seventy years of continuous encroachment substantially increases 'blue carbon' capacity as mangroves replace intertidal salt marshes. Global Change Biol 22:1097–1109

Kennedy H, Alongi DM, Karim A, Chen G, Chmura GL, Crooks S, Kairo JG, Liao B, Lin G, Troxler TG (2014) Coastal wetlands, chapter 4. In: Hiraishi T, Krug T, Tanabe K, Srivastava N, Jamsranjav B, Fukuda M, Troxler TG (eds) 2013 supplement to the 2006 IPCC guidelines for national greenhouse gas inventories: wetlands. IPCC, Gland

Kirwan ML, Megonigal JP (2013) Tidal wetland stability in the face of human impacts and sea-level rise. Nature 504:53–60

Kirwan ML, Mudd SM (2012) Response of salt-marsh carbon accumulation to climate change. Nature 489:550–554

Livesley SJ, Andrusiak SM (2012) Temperate mangrove and salt marsh sediments are a small methane and nitrous oxide source but important carbon store. Estuar Coast Shelf Sci 97:19–27

Loomis MJ, Craft CB (2010) Carbon sequestration and nutrient (nitrogen, phosphorus) accumulation in river-dominated tidal marshes, Georgia, USA. Soil Sci Soc Am J 74:1028–1036

Lovelock CE, Adame MF, Bennion V, Hayes M, O'Mara J, Reef R, Santini NS (2014) Contemporary rates of carbon sequestration through vertical accretion of sediments in mangrove forests and saltmarshes of South East Queensland, Australia. Estuar Coast 37:763–771

Macreadie PI, Nielsen DA, Kelleway JJ, Atwood TB, Seymour JR, Petrou K, Connolly RM, Thomson ACG, Trevathan-Tackett SM, Ralph PJ (2017a) Can we manage coastal ecosystems to sequester more blue carbon? Front Ecol Environ 15:206–213

Macreadie PI, Olliver QR, Kelleway JJ, Serrano O, Carnell PE, Lewis CJE, Atwood TB, Sanderman J, Baldock J, Connolly RM, Duarte CM, Lavery PS, Steven A, Lovelock CE (2017b) Carbon sequestration by Australian tidal marshes. Sci Rep 7:44071. https://doi.org/10.1038/srep44071

Markewich HW (1998) Carbon storage and late Holocene chronostratigraphy of a Mississippi River deltaic marsh, St. Bernard's Parish, Louisiana. No. 98-36. U.S. Geological Survey

McLeod E, Chmura GL, Bouillon S, Salm R, Björk M, Duarte CM, Lovelock CE, Schlesinger WH, Silliman BR (2011) A blueprint for blue carbon: toward an improved understanding of the role of vegetated coastal habitats in sequestering CO_2. Front Ecol Environ 9:552–560

Morris JT, Jensen A (1998) The carbon balance of grazed and non-grazed *Spartina anglica* saltmarsh at Skallingen, Denmark. J Ecol 86:229–242

Morris JT, Sundareshwar PV, Nietch CT, Kjerfve B, Cahoon DR (2002) Responses of coastal wetlands to rising sea level. Ecology 83:2869–2877

Nixon SW (1980) Between coastal marshes and coastal waters – a review of twenty years of speculation and research on the role of salt marshes in estuarine productivity and water chemistry. Springer, New York, pp 437–525

Orson RA, Warren RS, Niesing WA (1998) Interpreting sea-level rise and rates of vertical marsh accretion in a southern New England tidal salt marsh. Estuar Coast Shelf Sci 47:419–429

Ouyang X, Lee SY (2014) Updated estimates of carbon accumulation rates in coastal marsh sediments. Biogeosciences 11:5057–5071

Pendleton L, Donato DC, Murray BC, Crooks S, Jenkins WA, Sifleet S, Craft C, Fourqurean JW, Kauffman JB, Marbá N, Megonigal P, Pidgeon E, Herr D, Gordon D, Baldera A (2012) Estimating global "blue carbon" emissions from conversion and degradation of vegetated coastal ecosystems. PLoS One 7:e43542

Pethick JS (1981) Long-term accretion rates on tidal salt marshes. J Sediment Res 51:571–577

Robbins JA (1978) Geochemical and geophysical applications of radioactive lead. In: Nriagu JO (ed) Biogeochemistry of lead in the environment. Elsevier, Amsterdam, pp 285–393

Roman CT, Peck JA, Allen JR, King JW, Appleby DG (1997) Accretion of a New England salt marsh in response to inlet migration, storms and sea-level rise. Estuar Coast Shelf Sci 45:717–727

Saintilan N, Rogers K, Mazumder D, Woodroffe C (2013) Allochthonous and autochthonous contributions to carbon accumulation and carbon store in south eastern Australian coastal wetlands. Estuar Coast Shelf Sci 128:84–92

Schile LM, Kauffman JB, Crooks S, Fourqurean JW, Glavan J, Megonigal JP (2017) Limits on carbon sequestration in arid carbon ecosystems. Ecol Appl 27:859–874

Shepard CC, Caitlin CM, Beck MW (2011) The protective role of coastal marshes: a systematic review and meta-analysis. PLoS ONE 6:e27374

Sifleet S, Pendleton L, Murray BC (2011) State of the science on coastal blue carbon: a summary for policy makers. Nicolas Institute for Environmental Policy Solutions Report NI R 11-06. Nicolas Institute, Duke University, USA

Temmerman S, Govers G, Meir P, Wartel S (2003) Modelling long-term tidal marsh growth under changing conditions and suspended sediment concentrations, Scheldt estuary. Belgium. Mar Geol 212:151–169

Thomas CR, Blum LK (2010) Importance of fiddler crab *Uca pugnax* to salt marsh soil organic matter accumulation. Mar Ecol Prog Ser 414:167–177

Turner RE, Swenson EM, Milan CS (2001) Organic and inorganic contributions to vertical accretion in salt marsh sediments. In: Weinstein M, Kreeger K (eds) Concepts and controversies in tidal marsh ecology. Kluwer, Dordrecht, pp 583–595

Vernberg FJ (1993) Salt-marsh processes: a review. Environ Toxicol Chem 12:2167–2195

Wang J, Bai J, Zhao Q, Lu Q, Xia Z (2016) Five-year changes in soil organic carbon and total nitrogen in coastal wetlands affected by flow-sediment regulation in a Chinese delta. Sci Rept 6:21137

Weis P (2016) Salt marsh accretion. In: Kennish MJ (ed) Encyclopedia of estuaries. Springer, New York, pp 513–515

Weston NB, Neubauer SC, Velinsky DJ, Vile MA (2014) Net ecosystem carbon exchange and the greenhouse gas balance of tidal marshes along an estuarine salinity gradient. Biogeochem 120:163–189

Więski K, Guo H, Craft CB, Pennings SC (2010) Ecosystem functions of tidal marsh, brackish, and salt marshes on the Georgia coast. Estuar Coast 33:161–169

Ye S, Laws EA, Yuknis N, Ding X, Yuan H, Zhao G, Wang J, Yu X, Pei S, DeLaune RD (2015) Carbon sequestration and soil accretion in coastal wetland communities of the Yellow River delta and Liaohe delta, China. Estuar Coast 38:1885–1897

Yu J, Dong H, Li Y, Wu H, Guan B, Gao Y, Zhou D, Wang Y (2013) Spatiotemporal distribution characteristics of soil organic carbon in newborn coastal wetlands of the Yellow River delta estuary. Clean -Soil Air Water 42:311–316

Yuan J, Ding W, Liu D, Kang H, Freeman C, Xiang J, Lin Y (2014) Exotic *Spartina alterniflora* invasion alters ecosystem-atmosphere of CH_4 and N_2O and carbon sequestration in a coastal salt marsh in China. Global Change Biol 21:1567–1580

Zheng JF, Cheng K, Pan GX, Smith P, Li LQ, Zhang XH, Zheng JW, Han XJ, Du YL (2011) Perspectives on studies on soil carbon stocks and the carbon sequestration potential of China. Chinese Sci Bull 56:3748–3758

Zheng YM, Niu ZG, Gong P, Dai YJ, Shangguan W (2013) Preliminary estimation of the organic carbon pool in China's wetlands. Chinese Sci Bull 58:662–670

Chapter 3
Mangrove Forests

Mangrove forests are composed of woody trees and scrubs living along many coasts within low latitudes. These tidal forests attain peak luxuriance in sheltered muddy areas where quiescent conditions foster establishment and growth of propagules, but they do occur on rocky and sandy shores. Growing above mean sea-level, forest establishment involves positive feedback in which the trees trap silt and clay particles brought in by tides and rivers to help consolidate the deposits on which they grow. This feedback continues until the forest elevation lies above the reach of tides, and mangroves give way to terrestrial plants over years and decades (Alongi 2016).

As in salt marshes, the mangrove intertidal zone is highly dynamic in space and time, ever changing and disturbed often enough by storms and cyclones, disease, pests and anthropogenic intrusions that the natural progression from mangrove to land occurs rarely along most coastlines. Mangroves are subjected daily to a harsh environment, experiencing daily tides and seasonal variations in temperature, salinity and anoxic soils, and are thus highly robust and adaptable to ever-changing conditions.

Mangroves develop and persist in relation to the geomorphological evolution of low-latitude coastlines, pioneering newly formed mudflats but also shifting their intertidal position in the face of environmental change.

Mangrove development, like their salt marsh counterparts, can follow a number of patterns in relation to changes in sea-level. First, the mangrove surface may accrete asymptotically until sediment accumulation raises the forest floor above tidal range; this pattern occurs when sea-level is in equilibrium. Second, accretion may keep pace with a constant rise in sea-level. Third, the forest floor accretes at times above tidal range when sea-level rise is irregular. Fourth, the forest floor accretes back to the tidal range with episodic subsidence but with a stable sea-level. Fifth, under conditions of episodic subsidence but rising sea-level, mangrove accretion continues at an irregular pace. And finally, when there is no change in sediment volume with a rise in sea-level, the forest floor is set back (Woodroffe et al. 2016). These responses point to ever-changing conditions in which mangroves

© The Author(s) 2018
D. M. Alongi, *Blue Carbon*, SpringerBriefs in Climate Studies,
https://doi.org/10.1007/978-3-319-91698-9_3

have been traditionally classified as forests occupying overwash islands, coastal fringes, riverine areas and intertidal basins; scrub forests and other unique settings do occur including forests lying atop carbonate deposits as on coral islands.

Mangroves have evolved many morphological, reproductive and physiological traits for life in waterlogged saline soils including aerial roots, viviparous embryos, sclerophylly, low assimilation rates, high root/shoot ratios and high water-use and nutrient-use efficiencies. Forest structure is relatively simple compared with terrestrial forests, often lacking an understorey and having comparatively low tree diversity. Species richness is greatest in the Indo-West Pacific. Like salt marshes, tidal differences in species are frequently expressed in relation to combinations of tidal gradients in salinity, frequency of tidal inundation, seed predation, competition and other drivers, the complex interplay of which leads to forests that are mosaics of interrupted successional sequences.

There are about 70 true mangrove species in 40 genera in 25 families with 25 species in the families Rhizophoraceae and Avicenniaceae, plus a loosely defined group of mangrove associates that also occur in lowland rainforests, freshwater swamps and salt marshes. Mangrove food webs are dominated by bacteria and sesarmid and grapsid crabs but, like salt marsh food webs, have rich pelagic and benthic components consisting of both terrestrial and marine fauna and flora.

3.1   Capturing and Accumulating Sediment and Carbon

3.1.1   Mechanisms of Capture

Like salt marshes, mangroves actively and indirectly facilitate the capture and storage of sediment particles and associated carbon into soil horizons and capture sunlight to fuel growth and production of above- and below-ground biomass. Unlike salt marshes, above-ground biomass is substantial and can be an important store for carbon if left uncut. Mangroves are highly productive plants, and these forests can rival tropical rainforests in production and carbon storage, but can vary in size and age and thus in rates of primary productivity and carbon balance.

The dynamics of mangrove forests are similar to other forests in that there is an initial period of early rapid growth during colonisation with early establishment followed by a slow decline in growth rate into maturity and senescence. The mature old-growth phase is often prolonged such that an alternate succession state is reached as the climax stage is reset by successive disturbances. The net result of this phenomenon is that mangroves may be a carbon sink for up to a century if left relatively undisturbed.

Despite these capabilities, 75–95% of carbon in mangroves is stored as huge stocks below-ground in dead roots as most above-ground biomass is eventually lost due to clear cutting and human use, decomposition and export to the adjacent coastal zone (Donato et al. 2011; Alongi 2014). Over the long-term and under the right conditions, carbon is stored as peat. The accumulation of peat is a function not only of inputs from litter, roots, fallen tree stems and branches, algae and benthic organisms, such as burrowing crabs (Andreetta et al. 2014), but also slow decomposition rates of refractory material, the magnitude and frequency of tides, micro- and macro-organism activities, tree species and litter composition, moisture and temperature. As a result of a combination of these factors, peat formation and accumulation occur in some mangrove forests but not in others. Despite the fact that microbes and their enzymes are known to play a significant role in decomposition and accumulation of soil organic matter in mangroves, the underlying mechanisms of mangrove peat formation are not fully understood. Peat formation has been described as a 'enzyme latch' mechanism in which the amount of carbon storage is related to the inhibition of a single enzyme, phenol oxidase, under low oxygen conditions (Saraswati et al. 2016). This is in turn reported to result in the accumulation of phenolic materials which inhibits the activity of hydrolase enzymes which suppress the decomposition of organic matter, thus the term 'enzyme latch'. In laboratory experiments with peat from *Rhizophora mangle* forests, Saraswati et al. (2016) found that under aerobic conditions, soil samples have significantly higher phenol oxidase activity compared to anaerobic conditions. Soils supplemented with phenol oxidase show significantly lower phenolic concentration. These findings suggest that the 'enzyme latch' mechanism that operates in peatlands may also operate in mangrove peat soils.

As in salt marshes, carbon accumulation depends on a number of factors such as tidal amplitude, forest elevation, location in relation to the open coast and in relation to a tidal waterway, distance to adjacent aquatic habitats and primary productivity. Mangroves are not just passive importers of fine particulates but actively capture silt, clay and organic particles. Active capture involves maintaining particles in suspension in turbulent wakes created by tree trunks, prop roots and pneumatophores; most small flocs and free particles settle just before slack high tide. Despite the pull of ebb tide, most flocs and particles are retained within the forest as turbulence and water motion necessary for their resuspension is inhibited by the density of tree trunks. Due to the movement of the turbidity maximum zone where incoming bottom flow meets outward river flow within an estuary or waterway, mangrove waters have high suspended loads of mineral and organic particles. Tidal mixing, trapping and pumping within this zone facilitate flocculation and resuspension of particles. As these flocs and particles move into the forest on flood tide, turbulence generated by tidal flow around the trees helps to maintain flocs in suspension. The sticking of microbial mucus on the soil surface and the formation of excreted pellets by invertebrates facilitates rapid settling of particles.

The interrelationships between biotic and abiotic controls on soil accretion and elevation change as the same as those detailed in Sect. 2.1, as are the methods used to measure soil accretion. Woodroffe et al. (2016) reviewed the current status of knowledge of sedimentation and response of mangroves to relative sea-level rise and concluded that (1) accumulation rates of inorganic and organic, allochthonous and autochthonous sediment vary between and with environmental settings; (2) mangroves sequester carbon, but their sediments reveal paleo-environmental records of adjustments to past sea-level changes; (3) radiometric

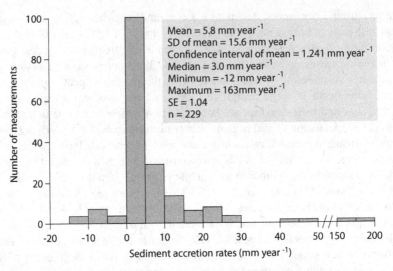

Fig. 3.1 Soil accretion rates measured in various mangrove forests worldwide. Updated from Alongi (2009, 2012) (Sources: Lynch et al. (1989); Furukawa and Wolanski (1996); Cahoon and Lynch (1997); Callaway et al. (1997); Alongi et al. (2004, 2005); Cahoon et al. (2003); Bird et al. (2004); Gonneea et al. (2004); Mahmood et al. (2005); Tateda et al. (2005); Whelan et al. (2005, 2009); Cahoon (2006); Rogers et al. (2006, 2013, 2014); McKee et al. (2007); Howe et al. (2009); Krauss et al. (2010); Alongi (2011); Sanders et al. (2010a, b, 2014); Stokes et al. (2010); Ceron-Breton et al. (2011); Lovelock et al. (2011a, b, 2015b); McKee (2011); Breithaupt et al. (2012); Oliver et al. (2012); Smoak et al. (2013); Lang'at et al. (2014); MacKenzie et al. (2016); Sasmito et al. (2016); Sidik et al. (2016); Ward et al. (2016); Hien et al. (2018); Pérez et al. (2018))

dating indicates long-term sedimentation, whereas RSET measurements indicate shallow subsurface processes of root growth and subsurface auto-compaction; (4) many tropical deltas also experience deep subsidence which augments relative sea-level rise; and (5) the persistence of mangroves implies an ability to cope with moderately high rates of relative sea-level rise. To persist, mangroves must build vertically at a rate equal to the combined rate of eustatic sea-level rise and land subsidence. Thus, mangroves have considerable natural resilience in response to sea-level (Krauss et al. 2013).

3.1.2   Rates of Soil Accretion and Carbon Sequestration

There has been an enormous growth in the literature of soil accretion rates in mangroves to the extent that a revised analysis of Alongi's (2012) figures is necessary. The rate of soil accretion in mangrove forests averages 5.8 mm year^{-1} with most measurements ranging from 0 to 2 mm year^{-1} (Fig. 3.1) The median is 3 mm year^{-1} with one standard error of 1.0 mm year^{-1} based on a sample size of $n = 229$.

A few measurements show either net erosion (Fig. 3.1) or massive accretion in highly impacted estuaries in China and Indonesia (Alongi et al. 2005; Sidik et al. 2016). Soil accretion rate is a function of tidal inundation frequency, as it is in salt marshes, as more frequent inundation of particle-laden water increases the frequency of particle settlement. Mangroves and salt marshes in high intertidal zones experience less soil accretion than wetlands located closer to the sea, so there is an overall pattern of decreasing sedimentation with decreasing tidal inundation frequency.

Below-ground roots and their ability to grow and vertically expand the soil are another driver of soil accretion, and surface growth of microbial mats and algae as well as litter and felled wood also contributes to vertical accretion. In some forests, these biotic forces can contribute more to vertical accretion than accumulation of particles via tides (McKee 2011; Krauss et al. 2013).

Natural subsidence plays a key role in long-term rates of soil accretion, being an important driver in estimating the susceptibility of mangroves to changes in sea-level (Woodroffe et al. 2016). Over long timescales, rates of vertical accretion vary in relation to climatic variability. Most mangroves are accreting sediment and carbon, but on some islands in the Pacific and in the Caribbean, sedimentation rates are slower than rates of sea-level rise. This is despite the fact that accretion rates on some of these islands are higher than eustatic sea-level rise (Sanders et al. 2010b). Throughout the Indo-Pacific, Lovelock et al. (2015c) found that recent trends indicate that at 69 percent of their study sites, the current rate of sea-level rise exceeds the soil accretion rate. They predict that sites with low tidal range and low sediment supply could be submerged as early as 2070. Sasmito et al. (2016) came to a similar conclusion that basin and fringe mangroves can keep pace with sea-level rise up to 2070 and 2055, respectively, on a global basis.

3.2 Carbon Sequestration Rates

The data ($n = 143$) for rates of carbon sequestration (CAR) in mangroves indicates an average (± 1 standard error) rate of 171 ± 17.1 g C_{org} m^{-2} year^{-1} with values ranging from 1 to 1053 g C_{org} m^{-2} year^{-1} with a median of 103 g C_{org} m^{-2} year^{-1} (Fig. 3.2). Assuming a global area of 137,760 km^2 (Giri et al. 2011) and using the median value, carbon sequestration in mangroves equates to 14.2 Tg C_{org} year^{-1}. This value is lower than the 23–25 Tg C_{org} year^{-1} calculated by Twilley et al. (1992), Jennerjahn and Ittekot (2002) and Duarte et al. (2005). Like the accretion data, the standard deviation (204 g C_{org} m^{-2} year^{-1}) is greater than the mean of 171 g C_{org} m^{-2} year^{-1} reflecting the high level of variability in carbon sequestration among mangroves of different ages and locations.

There is no clear relationship with differences in latitude as it is likely that these rates are a function of a number of interrelated factors such as forest age, tidal inundation frequency, tidal elevation, mangrove geomorphology, species composition, soil grain size, catchment and river input, ocean input and degree of human impact. Most values were in the range of 1–100 g C_{org} m^{-2} year^{-1} (half of all

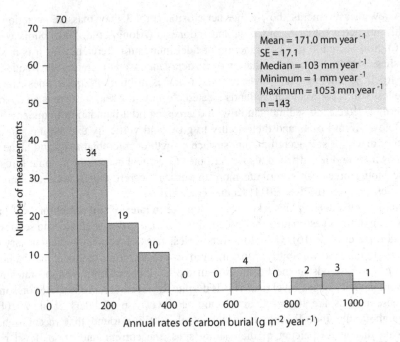

Fig. 3.2 Annual rates of carbon sequestration in various mangrove forests worldwide. Updated from Alongi (2012). (Sources: Lynch et al. (1989); Furakawa and Wolanski (1996); Callaway et al. (1997); Fujimoto et al. (1999); Alongi et al. (2004, 2005); Gonneea et al. (2004); Duarte et al. (2005); Mahmood et al. (2005); Tateda et al. (2005); Xiaonin et al. (2008); Alongi (2009, 2011); Ren et al. (2010); Sanders et al. (2010a, b, c, 2014); Ceron-Breton et al. (2011); Donato et al. (2011); Kauffman et al. (2011); McKee (2011); Ray et al. (2011, 2013); Mitra et al. (2011); Breithaupt et al. (2012); Matsui et al. (2012); Bianchi et al. (2013); Kathiresan et al. (2013); Smoak et al. (2013); Lunstrum and Chen (2014); Lovelock et al. (2015a); Zarate-Barrera and Maldonado (2015); Doughty et al. (2016); Ezcurra et al. (2016); MacKenzie et al. (2016); Marchio et al. (2016))

observations) with the highest values being from mature forests, those in close proximity to river deltas and forests in highly impacted catchments.

3.3   Carbon Stocks

Mangrove carbon stocks ($n = 168$) have been measured in 24 countries spanning the globe from the Americas to Africa to Asia (Table 3.1). Carbon stock for a mangrove forest averages 761.4 ± 45.5 Mg C_{org} ha^{-1} (± 1 SE) with a range of 37 to 2477 Mg C_{org} ha^{-1} and a median of 723.4 Mg C_{org} ha^{-1}. Using the median value and assuming a global mangrove area of 137,760 km^2, we derive a global carbon stock estimate for mangroves of 10 Pg. Jardine and Siilamäki (2014) estimated a global carbon stock of 5 Pg based on a predictive model using soil carbon concentrations against a high-resolution grid. They found that this stock is highly variable over space with considerable within-country variation.

At the forest level, the smallest carbon stocks are small stands that are primarily young plantations, while the largest stocks are mature stands. Because of the various ages and sizes of forest, there is no clear relationship with latitude; some equatorial forests are young stands, while some of the mature forests are at higher latitudes. Thus, older more mature forests store more carbon than young or scrub forests. The median value is close to the value of 703 Mg C ha^{-1} predicted by Jardine and Siilamäki (2014).

On average, 91.8% of the total ecosystem C_{org} stocks is vested below-ground (below-ground biomass + soil) with a mean above-ground to below-ground ratio of 11.2; the median value is 5.6 with a minimum value of 0.32 and a maximum of 83.2. The wide span of values reflects the wide range of ages and types of mangrove forest, from very young monocultures to mature forests. As with salt marshes, the average of 92% of carbon vested below-ground is a minimum estimate as many forests contain soil C_{org} stocks to depths greater than 1 m (Table 3.1).

3.4 Potential Losses

The carbon sequestration and carbon stock data suggest the potential for significant GHG emissions if the high per area carbon stocks of mangroves are disturbed. Losses of mangroves by clearing, conversion to industrial estates and aquaculture and changes in drainage patterns lead to dramatic changes in soil chemistry resulting in rapid emission rates of GHGs, especially CO_2. Lovelock et al. (2011a, b), for instance, measured the flux of CO_2 from mangrove peats that had been cleared for up to 20 years on the islands of Twin Cays in Belize and also measured gas effluxes after disturbing these cleared peats. They found that gas efflux declines from the time of first clearing from 10,600 tonnes km^{-2} year^{-1} in the first year to 3000 tonnes km^{-2} year^{-1} after 20 years since clearing; disturbing peats led to short-term increases in CO_2 efflux, but this returned to baseline levels within 2 days.

Using a stock-change approach, Kauffman et al. (2014) calculated that the potential emissions from the conversion of mangroves to shrimp ponds ranged from 2244 to 3799 Mg CO_2 equivalents ha^{-1} with all of the Dominican Republic's losses of mangroves estimated to have returned 3.8 Gg CO_2 equivalents or about 21% of the country's mangrove carbon stocks to conversion to the atmosphere, an amount that is among the largest measured carbon emissions from land use in the tropics. Kauffman et al. (2017) found that mangrove conversion to shrimp ponds results in GHG emissions ranging between 1067 and 3003 Mg CO_2 equivalents ha^{-1}, while conversion to cattle pastures results in losses estimated at 1464 Mg CO_2 equivalents ha^{-1} (Kauffman et al. 2016). Similarly, Murdiyarso et al. (2015) and Alongi et al. (2016) estimated that losses of Indonesian mangroves, marshes and seagrasses to conversion may equate to losses of roughly 29,040 Gg CO_2 equivalents to the atmosphere. In the world's largest continuous area of mangrove, the

Table 3.1 Estimates of organic carbon stocks (Mg C_{org} ha^{-1}) in mangrove biomass and soils to a depth of 1 m

Location [Sources]	Number of Observations	Range	Mean
Indonesia[a]	42	415–2202	1048
Vietnam[b]	18	979–1904	945
Honduras[c]	18	570–1060	921
United Arab Emirates[d]	18	77–515	218
India[e]	13	159–360	219
China[f]	11	114–619	321
Dominican Republic[g]	9	743–1142	922
Mexico[h]	8	381–1358	822
Ecuador[i]	6	425–580	485
Mozambique[j]	6	219–621	478
Ivory Coast[k]	4	51–176	128
Philippines[l]	4	241–660	438
Singapore[m]	4	37–227	133
Australia[e]	3	662–2139	1221
Malaysia[e]	3	995–1432	1267
Micronesia[n]	3	479–1218	1064
Palau[n]	3	625–840	720
Thailand[e]	3	579–808	662
Madagascar[o]	3	367–593	499
Bangladesh[p]	2	343–604	566
Cameroon[q]	2	2102–2477	2289
Japan[r]	1		107
Myanmar[s]	1		274
Colombia[t]	1		196
USA[u]	1		122
Senegal[v]	1		674
Liberia[v]	1		949
Gabon[v]	1		801

[a]Sources: Donato et al. (2011) and Murdiyarso et al. (2015)
[b]Sources: Alongi (2012) and Nam et al. (2016)
[c]Source: Bhomia et al. (2016a)
[d]Source: Schiele et al. (2017)
[e]Source: Rahman et al. (2015) and Bhomia et al. (2016b)
[f]Sources: Alongi (2012), Alongi (unpublished data), Lu et al. (2014), and Lunstrum and Chen (2014)
[g]Source: Kauffman et al. (2014)
[h]Source: Adame et al. (2013, 2015a, b) and Kauffman et al. (2016)
[i]Source: DelVecchia et al. (2014)
[j]Sources: Sitoe et al. (2014) and Stringer et al. (2015)
[k]Source: Osemwegie et al. (2016)
[l]Sources: Thompson et al. (2014) and Bigsang et al. (2016)
[m]Source: Friess et al. (2016)
[n]Source: Kauffman et al. (2011) and Donato et al. (2012)
[o]Source: Jones et al. (2014, 2015)
[p]Source: Donato et al. (2011)
[q]Source: Ndema et al. (2016)
[r]Source: Khan et al. (2007)
[s]Source: Thant et al. (2012)
[t]Source: Zarate-Barrera and Maldonado (2015)
[u]Source: Doughty et al. (2016)
[v]Source: Kauffman et al. (2017)

Sundarbans of India, Akhand et al. (2016) estimated that between 1975 and 2013, potential carbon dioxide emission due to the degradation of just the above-ground biomass of mangroves was about 1570 Gg. Globally, Pendleton et al. (2012) estimated that total loss of mangroves may account for about 0.09 to 0.45 Pg CO_2 equivalents year^{-1}.

Using the known area of mangroves (137,760 km^2; Giri et al. 2011) and the median carbon stock (723.4 Mg C_{org} ha^{-1}) and assuming a destruction rate of 1–2% per year, we can estimate a loss of between 0.27 and 0.59 Pg CO_2 equivalents year^{-1} which is within the wide range estimated by Pendleton et al. (2012) and is an order of magnitude greater than the estimate of Atwood et al. (2017). The annual losses of mangroves add another 5–11% to the recent estimate (Hansen et al. 2013) of global deforestation (4.6 Pg CO_2 year^{-1}) or offset 23–49% of the carbon sink in the global ocean's continental margins (Chen and Borges 2009). These are only rough estimates, but the range of values underscores the global significance of continuing mangrove losses. If all of the world's mangrove forests were destroyed and assuming that 95% of all mangrove carbon was oxidised to CO_2 (Kennedy et al. 2014), the loss would be 30.2 Pg CO_2 equivalents which is equal to 6.5 years of carbon emissions from global forest loss.

References

Adame MF, Kauffman JB, Medina I, Gamboa JN, Torres O, Caamal JP, Reza M, Herrera-Silveira JA (2013) Carbon stocks of tropical coastal wetlands within the karstic landscape of the Mexican Caribbean. PLoS ONE 8:e56569

Adame MF, Hermoso V, Perhans K, Lovelock CE, Herrerea-Silveira JA (2015a) Selecting cost-effective areas for restoration of ecosystem services. Conserv Biol 29:493–502

Adame MF, Santini NS, Tovilla C, Vázquez-Lule A, Castro L (2015b) Carbon stocks and soil sequestration rates in riverine mangroves and freshwater wetlands. Biogeosci Discuss 12:1015–1045

Akhand A, Mukhopadhyay A, Chanda A, Mukherjee S, Das A, Das S, Hazra S, Mitra D, Choudhury SB, Rao KH (2016) Potential CO_2 emission due to loss of above ground biomass from the Indian Sundarban mangroves during the last four decades. J Indian Soc Remote Sens 8:1–8

Alongi DM (2009) The energetics of mangrove forests. Springer, Amsterdam

Alongi DM (2011) Carbon payments for mangrove conservation: ecosystem constraints and uncertainties of sequestration potential. Environ Sci Policy 14:462–470

Alongi DM (2012) Carbon sequestration in mangrove forests. Carbon Manage 3:313–322

Alongi DM (2014) Carbon cycling and storage in mangrove forests. Annu Rev Mar Sci 6:195–219

Alongi DM (2016) Mangroves. In: Kennish MJ (ed) Encyclopedia of estuaries. Springer, New York, pp 393–404

Alongi DM, Sasekumar A, Chong VC, Pfitzner J, Trott LA, Tirendi F, Dixon P, Brunskill GJ (2004) Sediment accumulation and organic material flux in a managed mangrove ecosystem: estimates of land-ocean-atmosphere exchange in peninsular Malaysia. Mar Geol 208:383–402

Alongi DM, Pfitzner J, Trott LA, Tirendi F, Dixon P, Klumpp DW (2005) Rapid sediment accumulation and microbial mineralization in forests of the mangrove *Kandelia candel* in the Jiulongjiang estuary, China. Estuar Coast Shelf Sci 63:605–618

Alongi DM, Murdiyarso D, Fourqurean JW, Kauffman JB, Hutahaean A, Crooks S, Lovelock CE, Howard J, Herr D, Fortes M, Pidgeon E, Wagey T (2016) Indonesia's blue carbon: a globally significant and vulnerable sink for seagrass and mangrove carbon. Wetl Ecol Manage 24:3–13

Andreetta A, Fusi M, Cameldi I, Cimó F, Carnicelli S, Canicci S (2014) Mangrove carbon sink. Do burrowing crabs contribute to sediment carbon storage? Evidence from a Kenyan mangrove system. J Sea Res 85:524–533

Atwood TB, Connolly RM, Almahasheer H, Carnell PE, Duarte CM, Lewis CJE, Irigoien X, Kelleway JJ, Lavery PS, Macreadie PI, Serrano O, Sanders CJ, Santos I, Steven ADL (2017) Global patterns in mangrove soil carbon stocks and losses. Nat Clim Chang 7:523–528. https://doi.org/10.1038/NCLIMATE3326

Bhomia RK, Kauffman JB, McFadden TN (2016a) Ecosystem carbon stocks of mangrove forests along the Pacific and Caribbean coasts of Honduras. Wetl Ecol Manage 24:187–201

Bhomia RK, Mackenzie RA, Murdiyarso D, Sasmito SD, Purbopuspito J (2016b) Impacts of land-use on Indian mangrove forest carbon stocks: implications for conservation and management. Ecol Appl 26:1396–1408

Bianchi TS, Allison MA, Zhao J, Li X, Comeaux RS, Feagin RA, Kulawardhana RW (2013) Historical reconstruction of mangrove expansion in the Gulf of Mexico: linking climate change with carbon sequestration in coastal wetlands. Estuar Coast Shelf Sci 119:7–16

Bigsang RT, Agonia NB, Toreta CGD, Nacin CJCB, Obemio CDG, Martin TTB (2016) Community structure and carbon sequestration potential of mangroves in Maasim, Sarangani Province, Philippines. Adv Environ Sci 8:6–13

Bird MI, Fifield LK, Chua S, Goh B (2004) Calculating sediment compaction for radiocarbon dating of intertidal sediments. Radiocarbon 46:421–436

Breithaupt JL, Smoak JM, Smith TJ III, Sanders CJ, Hoare A (2012) Organic carbon burial rates in mangrove sediments: strengthening the global budget. Global Biogeochem Cycles 26:GB3011

Cahoon DR (2006) A review of major storm impacts on coastal wetland elevations. Estuar Coasts 29:889–898

Cahoon DR, Lynch JC (1997) Vertical accretion and shallow subsidence in a mangrove forest of southwestern Florida, USA. Mangroves Salt Marshes 1:173–186

Cahoon DR, Hensel P, Rybczyz J, McKee KL, Proffitt E, Perez BC (2003) Mass tree mortality leads to mangrove peat collapse at Bay Islands, Honduras after Hurricane Mitch. J Ecol 91:1093–1105

Callaway JC, DeLaune RD, Patrick WH Jr (1997) Sediment accretion rates from four coastal wetlands along the Gulf of Mexico. J Coast Res 13:181–191

Cerón-Bretón JG, Cerón-Bretón RM, Rangel-Marrón M, Muriel-García M, Cordova-Quiroz AV, Estrella-Cahuich A (2011) Determination of carbon sequestration rate in soil of a mangrove forest in Campeche, Mexico. WSEAS Trans Environ Devel 7:55–64

Chen CTA, Borges AV (2009) Reconciling opposing views on carbon cycling in the coastal ocean: continental shelves as sinks and near-shore ecosystems as sources of atmospheric CO_2. Deep-Sea Res II 56:578–590

DelVecchia AG, Bruno JF, Benninger L, Alperin M, Banerjee O, Morales JD (2014) Organic carbon inventories in natural and restored Ecuadorian mangrove forests. Peer J 2:e388

Donato DC, Kauffman JB, Murdiyarso D, Kurnianto S, Stidham M, Kanninen M (2011) Mangroves among the most carbon-rich forests in the tropics. Nature Geosci 4:293–297

Donato DC, Kauffman JB, Mackenzie RA, Ainsworth A, Pfleeger AZ (2012) Whole-island carbon stocks in the tropical Pacific: implications for mangrove conservation and upland restoration. J Environ Manage 97:89–96

Doughty CL, Langley JA, Walker WS, Feller IC, Schaub R, Chapman SK (2016) Mangrove range expansion rapidly increases coastal wetland carbon storage. Estuar Coasts 39:385–396

Duarte CM, Middelburg JJ, Caraco N (2005) Major role of marine vegetation on the oceanic carbon cycle. Biogeosciences 2:1–8

Ezcurra P, Ezcurra E, Garcillán PP, Costa MT, Aburto-Oropeza O (2016) Coastal landforms and accumulation of mangrove peat increase carbon sequestration and storage. Proc Nat Acad Sci USA 113:4404–4409

Friess DA, Richards DR, Phang VXH (2016) Mangrove forests store high densities of carbon across the tropical urban landscape of Singapore. Urban Ecosyst 19:795–810

Fujimoto K, Imaza A, Tabuchi R, Kuramoto S, Utsugi H, Murofushi T (1999) Belowground carbon storage of Micronesian mangrove forests. Ecol Res 14:409–413

Furukawa K, Wolanski E (1996) Sedimentation in mangrove forests. Mangroves Salt Marshes 1:3–10

Giri C, Ochieng E, Tiezen LL, Zhu Z, Singh A, Loveland T, Masek J, Duke NC (2011) Status and distribution of mangrove forests of the world using earth observation satellite data. Global Ecol Biogeogr 20:154–159

Gonneea ME, Paytan A, Herrera-Silveira JA (2004) Tracing organic matter sources and carbon burial in mangrove sediments over the past 160 years. Estuar Coast Shelf Sci 61:211–227

Hansen MC, Potapov PV, Moore R, Hancher M, Turubanova TA, Thau D, Stehman SV, Goetz SJ, Loveland TR, Kommareddy A, Egorov A, Chini L, Justice CO, Townshend JRG (2013) High-resolution global maps of 21st-century forest cover change. Science 342:850–853

Hien HT, Marchand C, Aimé J, Nhon DH, Hong PN, Tung NX, Cuc NTK (2018) Belowground carbon sequestration in a mature planted mangroves (Northern Viet Nam). Forest Ecol Manag 407:191–199

Howe AJ, Rodriguez JF, Saco PM (2009) Surface evolution and carbon sequestration in disturbed and undisturbed wetland soils of the Hunter estuary, southeast Australia. Estuar Coast Shelf Sci 84:75–83

Jardine SL, Siikamäki JV (2014) A global predictive model of carbon in mangrove soils. Environ Res Lett 9:104013

Jennerjahn TC, Ittekot V (2002) Relevance of mangroves for the production and deposition of organic matter along tropical continental margins. Naturwissenschaften 89:23–30

Jones TG, Ratsimba HR, Ravaoarinorotsihoarana L, Cripps G, Bey A (2014) Ecological variability and carbon stock estimates of mangrove ecosystems in north western Madagascar. Forests 5:177–205

Jones TG, Ratsimba HR, Ravaoarinorotsihoarana L, Glass L, Benson L, Teoh M, Carro A, Cripps G, Giri C, Gandhi S, Andriamahenina Z, Rakotomanana R, Roy P-F (2015) The dynamics, ecological variability and estimated carbon stocks of mangroves in Mahajamba Bay, Madagascar. J Mar Sci Eng 3:793–820

Kathiresan K, Anburaj R, Gomathi V, Saravanakumar K (2013) Carbon sequestration potential of *Rhizophora mucronata* and *Avicennia marina* as influenced by age, season, growth and sediment characteristics in southeast coast of India. J Coast Conserv 17:397–408

Kauffman JB, Heider C, Cole TG, Dwire KA, Donato DC (2011) Ecosystem carbon stocks of Micronesian mangrove forests. Wetlands 31:343–352

Kauffman JB, Heider C, Norfolk J, Payton F (2014) Carbon stocks of intact mangroves and carbon emissions arising from their conversion in the Dominican Republic. Ecol Appl 24(24):518–527

Kauffman JB, Trejo HH, Garcia MCJ, Heider C, Contreras WM (2016) Carbon stocks of mangroves and losses arising from their conversion to cattle pastures in the Pantanos de Centla, Mexico. Wetl Ecol Manag 24:203–216

Kauffman JB, Arifanti VB, Trejo HH, Garcia MCJ, Norfolk J, Cifuentes M, Hadriyanto D, Murdiyarso D (2017) The jumbo carbon footprint of a shrimp: carbon losses from mangrove deforestation. Front Ecol Environ. https://doi.org/10.1002/fee.1482

Kennedy H, Alongi DM, Karim A, Chen G, Chmura GL, Crooks S, Kairo JG, Liao B, Lin G, Troxler TG (2014) Coastal wetlands, chapter 4. In: Hiraishi T, Krug T, Tanabe K, Srivastava N, Jamsranjav B, Fukuda M, Troxler TG (eds) 2013 supplement to the 2006 IPCC guidelines for national greenhouse gas inventories: wetlands. IPCC, Gland

Khan Md NI, Suwa R, Hagihara A (2007) Carbon and nitrogen pools in a mangrove stand of *Kandelia obovate* (S.L.) Yong: vertical distribution in the soil vegetation system. Wetl Ecol Manage 15:141–153

Krauss KW, Cahoon DR, Allen JA Ewel KC, Lynch JC, Cormier N (2010) Surface elevation change and susceptibility of different mangrove zones to sea-level rise on Pacific high islands of Micronesia. Ecosystems 13:129–143

Krauss KW, McKee KL, Lovelock CE, Cahoon DR, Saintilan N, Reef R, Chen L (2013) How mangrove forests adjust to rising sea level. New Phytol 202:19–34

Lang'at JKS, Kairo JG, Mencuccini M (2014) Rapid losses of surface elevation following tree girdling and cutting in tropical mangroves. PLoS One 9:e107868

Lovelock CE, Feller IC, Adame M, Reef R, Penrose H, Wei L, Ball MC (2011a) Intense storms and the delivery of materials that relieve nutrient limitation in mangroves of an arid zone estuary. Funct Plant Biol 38:514–522

Lovelock CE, Ruess RW, Feller IC (2011b) CO_2 efflux from cleared mangrove peat. PLoS ONE 6: e21279

Lovelock CE, Simpson LT, Duckett LJ, Feller IC (2015a) Carbon budgets for Caribbean mangrove forests of varying structure and with phosphorus enrichment. Forests 6:3528–3546

Lovelock CE, Adame MF, Bennion V, Hayes M, Reef R, Santini N, Cahoon DR (2015b) Sea level and turbidity controls on mangrove soil surface elevation change. Estuar Coast Shelf Sci 153:1–9

Lovelock CE, Cahoon DR, Friess DA, Guntenspergen GR, Krauss KW, Reef R, Rogers K, Saunders ML, Sidik F, Swales A, Saintilan N, Thuyen LX, Triet T (2015c) The vulnerability of Indo-Pacific mangrove forests to sea-level rise. Nature 526:559–563

Lu W, Yang S, Chen L, Wang W, Du X, Wang C, Ma Y, Lin G, Lin G (2014) Changes in carbon pool and stand structure of a native subtropical mangrove forest after inter-planting with exotic species *Sonneratia apetala*. PLoS ONE 9:e91238

Lunstrum A, Chen L (2014) Soil carbon stocks and accumulation in young mangrove forests. Soil Biol Biochem 75:223–232

Lynch JC, Meriwether JR, McKee BA, Vera-Herrera F, Twilley RR (1989) Recent accretion in mangrove ecosystems based on ^{137}Cs and [210]Pb. Estuaries 12:284–299

MacKenzie RA, Foulk PB, Val Klump J, Weckerly K, Purbospito J, Murdiyarso D, Donato DC, Nam VN (2016) Sedimentation and belowground carbon accumulation rates in mangrove forests that differ in diversity and land use: a tale of two mangroves. Wetl Ecol Manage 24:245–261

Mahmood H, Misri K, Japar Sidik B, Saberi O (2005) Sediment accretion in a protected mangrove forest of Kuala Selangor, Malaysia. Pak J Biol Sci 8:149–151

Marchio DA, Savarese M, Bovard B, Mitch WJ (2016) Carbon sequestration and sedimentation in mangrove swamps influenced by hydrogeomorphic conditions and urbanization in Southwest Florida. Forests 7:116. https://doi.org/10.3390/f7060116

Matsui N, Morimune K, Meepol W, Chukwamdee J (2012) Ten-year evaluation of carbon stock in mangrove planation reforested from an abandoned shrimp pond. Forests 3:431–444

McKee KL (2011) Biophysical controls on accretion and elevation change in Caribbean mangrove ecosystems. Estuar Coast Shelf Sci 91:475–483

McKee KL, Cahoon DR, Feller IC (2007) Caribbean mangroves adjust to rising sea level through biotic controls on change in soil elevation. Glob Ecol Biogeogr 16:545–556

Mitra A, Sengupta K, Banerjee K (2011) Standing biomass and carbon storage of above-ground structures in dominant mangrove trees in the Sundarbans. Forest Ecol Manage 261:1325–1335

Murdiyarso D, Purbopuspito J, Kauffman JB, Warren MW, Sasmito SD, Donato DC, Manuri S, Krisnawati H, Taberima S, Kurnianto S (2015) The potential of Indonesian mangrove forests for global climate change mitigation. Nature Clim Change 5:1089–1092

Nam VN, Sasmito SD, Murdiyarso D, Purbopuspito J, MacKenzie RA (2016) Carbon stocks in artificially and naturally regenerated mangrove ecosystems in the Mekong Delta. Wetl Ecol Manage 24:231–244

Ndema Nsombo E, Ako'o Bengono F, Etame J, Din N, Ajonina G, Bilong P (2016) Effects of vegetation's degradation on carbon stock. Morphological, physical and chemical characteristics of soils within the mangrove forest of the Rio del Rey estuary: case study-Bamusso (South-West Cameroon). Afr J Environ Sci Tech 10:58–66

Oliver TSN, Rogers K, Chafer CJ, Woodroffe CD (2012) Measuring, mapping and modelling: an integrated approach to the management of mangrove and saltmarsh in the Minnamurra estuary, southeast Australia. Wetl Ecol Mange 20:353–371

Osemwegie I, DN'da H, Stumpp C, Reichart B, Biemi J (2016) Mangrove forest characterization in Southeast Côte d'Ivoire. Open J Ecol 6:138–150

Pendleton L, Donato DC, Murray BC, Crooks S, Jenkins WA, Sifleet S, Craft C, Fourqurean JW, Kauffman JB, Marbá N, Megonigal P, Pidgeon E, Herr D, Gordon D, Baldera A (2012) Estimating global "blue carbon" emissions from conversion and degradation of vegetated coastal ecosystems. PLoS One 7:e43542

Pérez A, Machado W, Gutiérrez D, Borges AC, Patchineelam SR, Sanders CJ (2018) Carbon accumulation and storage capacity in mangrove sediments three decades after deforestation within a eutrophic bay. Mar Pollut Bull 126:275–280

Rahman MM, Khan MNI, Hoque AKF, Ahmed I (2015) Carbon stock in the Sundarbans mangrove forest: spatial variations in vegetation types and salinity zones. Wetl Ecol Manage 23:269–283

Ray R, Ganguly D, Chowdhury C, Dey M, Das S, Dutta MK, Mandal SK, Majumber N, De TK, Mukhopadhyay SK, Jana TK (2011) Carbon sequestration and annual increase of carbon stock in a mangrove forest. Atmos Environ 45:5016–5024

Ray R, Chowdhury C, Majumder N, Dutta MK, Mukhopadhyay SK, Jana TK (2013) Improved model calculation of atmospheric CO$_2$ increment in affecting carbon stock of tropical mangrove forest. Tellus B 65:18981

Ren H, Chen H, Li Z, Han W (2010) Biomass accumulation and carbon storage of four different aged *Sonneratia apetala* plantations in Southern China. Plant Soil 327:279–291

Rogers K, Wilton K, Saintilan N (2006) Vegetation change and surface elevation dynamics in estuarine wetlands of southeast Australia. Estuar Coast Shelf Sci 66:559–569

Rogers K, Saintilan N, Howe A, Rodriquez J (2013) Sedimentation, elevation and marsh evolution in a southeastern Australian estuary during changing climatic conditions. Estuar Coast Shelf Sci 133:172–181

Rogers K, Saintilan N, Woodroffe C (2014) Surface elevation changes and vegetation distribution dynamics in a subtropical coastal wetland: implications for coastal wetland response to climate change. Estuar Coast Shelf Sci 149:46–56

Sanders CJ, Smoak JM, Naidu AS, Araripe DR, Sanders LM, Patchineelam SR (2010a) Mangrove forest sedimentation and its reference to sea-level rise. Environ Earth Sci 60:1291–1301

Sanders CJ, Smoak JM, Naidu AS, Sanders LM, Patchineelam SR (2010b) Organic carbon burial in a mangrove forest, margin and intertidal mud flat. Estuar Coast Shelf Sci 90:169–172

Sanders CJ, Smoak JM, Sanders LM, Naidu AS, Patchineelam SR (2010c) Organic carbon accumulation in Brazilian mangal sediments. J South Am Earth Sci 30:189–192

Sanders CJ, Eyre BD, Santos IR, Machado W, Luiz-Silva W, Smoak JM, Breithaupt JL, Ketterer ME, Sanders L, Marotta H, Silva-Filho E (2014) Elevated rates of organic carbon, nitrogen, and phosphorus accumulation in a highly impacted mangrove wetland. Geophys Res Lett 41:2475–2480

Saraswati S, Dunn C, Mitsch WJ, Freeman C (2016) Is peat accumulation in mangrove swamps influenced by the 'enzyme latch' mechanism? Wetl Ecol Manage 24:641–650

Sasmito SD, Murdiyarso D, Friess DA, Kurnianto S (2016) Can mangroves keep pace with contemporary sea level rise? A global data review. Wetl Ecol Manage 24:263–278

Schile LM, Kauffman JB, Crooks S, Fourqurean JW, Glavan J, Megonigal JP (2017) Limits on carbon sequestration in arid carbon ecosystems. Ecol Appl 27:859–874

Sidik F, Neil D, Lovelock CE (2016) Effect of high sedimentation rates on surface sediment dynamics and mangrove growth in the Porong River, Indonesia. Mar Pollut Bull 107:355–363

Sitoe AA, Mandlate LJC, Guesdes BS (2014) Biomass and carbon stocks of Sofala Bay mangrove forests. Forests 5:1967–1981

Smoak JM, Breithaupt JL, Smith TJ III, Sanders CJ (2013) Sediment accretion and organic carbon burial relative to sea-level rise and storm events in two mangrove forests in Everglades National Park. Catena 104:58–66

Stokes DJ, Healy TR, Cooke PJ (2010) Expansion dynamics of monospecific, temperate mangroves and sedimentation in two embayments of a barrier-enclosed lagoon, Tauranga Harbour, New Zealand. J Coast Res 261:113–122

Stringer CE, Trettin CC, Zarnoch SJ, Tang W (2015) Carbon stocks of mangroves within the Zambezi River delta, Mozambique. For Ecol Manag 354:139–148

Tateda Y, Nhan DD, Wattayakorn G, Toriumi H (2005) Preliminary evaluation of organic carbon sedimentation rates in Asian mangrove coastal ecosystems estimated by ^{210}Pb chronology. Radioprotection 40:S527–S532

Thant YM, Kanzaki M, Ohta S, Than MM (2012) Carbon sequestration by mangrove plantations and a natural regeneration stand in the Ayeyarwady delta, Myanmar. Tropics 21:1–10

Thompson BS, Clubbe CP, Primavera JH, Curnick D, Koldewey HJ (2014) Locally assessing the economic viability of blue carbon: a case study from Panay Island, the Philippines. Ecosyst Serv 8:128–140

Twilley RR, Chen RH, Hargis T (1992) Carbon sinks in mangroves and their implications to carbon budget of tropical coastal ecosystems. Water Air Soil Pollut 64:265–288

Ward RD, Friess DA, Day RH, MacKenzie RA (2016) Impacts of climate change on mangrove ecosystems: a region by region overview. Ecosyst Health Sustain 2:e01211

Whelan KRT, Smith TJ III, Cahoon DR, Lynch JC, Anderson GH (2005) Groundwater control of mangrove surface elevation: shrink and swell varies with soil depth. Estuaries 28:833–843

Whelan KRT, Smith TJ III, Anderson GH, Ouellette ML (2009) Hurricane Wilma's impact on overall soil elevation and zones within the soil profile in a mangrove forest. Wetlands 29:16–23

Woodroffe CD, Rogers K, McKee KL, Lovelock CE, Mendelssohn IA, Saintilan N (2016) Mangrove sedimentation and response to relative sea-level rise. Annu Rev Mar Sci 8:243–266

Xiaonin D, Xiake W, Lu F, Zhiyun O (2008) Primary evaluation of carbon sequestration potential of wetlands in China. Acta Ecol Sinica 28:463–469

Zarate-Barrera TG, Maldonado JH (2015) Valuing blue carbon: carbon sequestration benefits provided by the marine protected areas in Colombia. PLoS ONE 10:e0126627

Chapter 4
Seagrass Meadows

Seagrass meadows are intertidal and shallow subtidal habitats composed of up to 76 species of marine angiosperms and are important components of global estuarine and coastal ecosystems in boreal, temperate and tropical latitudes. Found on all continents except Antarctica, seagrasses provide habitat, protection and nursery grounds for economically valuable fishery species, act as indicators of and modify local water quality and form close links between benthic and pelagic food chains, and nutrient and carbon cycles (Jackson et al. 2001; Mateo et al. 2006; Unsworth et al. 2014). They have a high level of connectivity with mangroves and coral reefs (Unsworth et al. 2008) and are important habitats for food security and human well-being (Cullen-Unsworth et al. 2014).

Seagrasses are among the most productive primary producers in the sea and, like mangroves and salt marshes, have strong trophic links to the coastal ocean (Holmer 2009). Roughly half of their primary productivity is contributed by the seagrasses themselves with the other half coming from associated epiphytes and macroalgae. In tropical areas where seagrass species diversity (up to 12 species) is greater than in higher latitudes, dugong, sea turtles and parrotfish directly feed on these angiosperms. Many tropical seagrass species are highly productive to the extent that they can provide most of the fixed carbon for some coastal regions.

A large fraction of this fixed carbon is not consumed by herbivores, and seagrass tissue is relatively refractory and decomposes slowly. A significant fraction of seagrass production occurs below-ground as roots and rhizomes where this material can be preserved over long time scales (Duarte et al. 2005). Seagrass meadows are net autotrophic, acting as net CO_2 sinks. Until recently, the role of seagrasses in storing carbon has been ignored.

Like salt marshes and mangroves, seagrass meadows are highly dynamic in time and space with large changes taking place over short intervals. Physical disturbance, herbivory, intraspecific competition, nutrients, pollution and deposition of fine particles all play key roles in influencing seagrass biomass, species composition and area. A number of factors will determine if seagrasses will occur in any given area, including natural biophysical drivers that regulate physiological activity

© The Author(s) 2018
D. M. Alongi, *Blue Carbon*, SpringerBriefs in Climate Studies,
https://doi.org/10.1007/978-3-319-91698-9_4

and morphology, such as light availability, temperature, water clarity, salinity, wave action, currents, depth, substrate, day length, nutrients, epiphytes and diseases. Also, the availability of seeds and vegetative fragments and anthropogenic inputs, such as sediment loading and excess nutrients may be important determinants in seagrass existence.

Widespread losses of seagrass have occurred globally, and about 24% of all species are at risk of extinction or are now classified as near threatened on the IUCN's Red List (Waycott et al. 2009; Short et al. 2011). The rate of seagrass decline has increased over the past 70 years, from 0.9% per year prior to 1940 to 7% per year since 1980. Direct impacts such as removal of seagrass during dredging cause immediate loss, but a large number of indirect impacts cause much of the permanent and chronic damage to seagrass meadows. These include overfishing, long-term nutrient pollution and climate change.

Few metabolic studies have been conducted in the Southern Hemisphere to investigate whether or not seagrass meadows have potential as carbon sinks (Duarte et al. 2010), but the few studies available indicate that they have large storage capacity (Duarte et al. 2011) and can form the basis for climate change mitigation strategies. Seagrass meadows function to trap and bind sediment by trapping suspended particles from currents and hereby help to clarify the overlying water column. The root and rhizomes stabilise the sediments and help prevent coastal erosion during storms, heavy rains and floods. Seagrass detritus is not only an important trophic link, but accumulates to become an important carbon sink.

4.1 Fluid Dynamics: The Mechanism for Sediment and Carbon Accumulation

Seagrasses, like their salt marsh and mangrove counterparts, are ecosystem engineers capable by their very existence of reducing the velocity of currents and attenuating waves to the extent that sediment particles can deposit on their surfaces and on the seabed. Other factors play important roles in helping to accumulate carbon, such as canopy complexity, turbidity, wave height and water depth (Samper-Villarreal et al. 2016). But the essence of what drives the accumulation of sediment particles and associated carbon is fluid dynamics. The movement of water among, between and around seagrass blades is the key feature of carbon capture (Koch et al. 2006).

The main source of energy required to move water is the sun which causes winds that lead to waves and thermal gradients that lead to expansion, mixing and instabilities in water gradients and thus flow. Seawater, being an incompressible fluid, moves at a flow rate (Q) which is defined by the velocity (u) of the fluid that passes through a cross-sectional area, A. Water flow leads to both hydrostatic and dynamic pressures which are a constant. What this means in practical terms is that the sum of the pressures helps to explain lift that occurs within, around and under seagrass canopies. Drag is another force that operates in the case of water motion and has two components, (1) viscous drag (F_d) that exists due to the interaction of the seagrass surface with the water and defined as

$$F_d = 1/2C_d\rho Au^2 \qquad (4.1)$$

where C_d is the drag coefficient and ρ is the hydrostatic pressure and (2) the dynamic or pressure drag (F_p) that exists under high flows when flows separate from boundaries.

Water flow can be either smooth and regular (laminar flow) or rough and irregular (turbulent flow), depending on the velocity and temporal and spatial scale under investigation as defined by the Reynolds number:

$$R_e = lu/v \qquad (4.2)$$

where l is the length scale under observation and v is the kinematic viscosity. R_e defines four flow regimes that may occur: (1) creeping flow where $R_e \ll 1$ which occurs at very slow flows and spatial scales such as those experienced by microbes, (2) laminar flow ($1 < R_e < 10^3$) which is smooth and regular, (3) transitional flow ($R_e \approx 10^3$) which involves the production of eddies and disturbances in the flow and (4) fully turbulent flow ($R_e \gg 3$). These flows are scale-dependent; flow is almost always turbulent across entire seagrass meadows but laminar at the scale of individual seagrass leaves.

Flow conditions become more complex when water approaches a boundary such as the seagrass canopy or seafloor. Water cannot penetrate such boundaries but slips by it, a condition which leads to the development of a velocity gradient perpendicular to the boundary as the velocity at the boundary will be zero relative to the stream velocity (U_0). As water flows downstream, the velocity gradient will get larger and a slower moving layer of water will develop next to the boundary. Vertically, there is a sublayer in which the forces are largely viscous. Consequently, the mass transfer in this layer is slow, dominated by diffusion, which is called a diffusive boundary layer. Such boundary layers can become embedded within one another such that it is possible to define boundary conditions around blade epiphytes, flowers, leaves and the canopy.

At the molecular level, a boundary layer develops on the sediment surface as well as on each leaf, shoot or flower as water flows through a seagrass meadow. The faster the water movement, the thinner the diffusive boundary layer, and thus the transfer of molecules (e.g. CO_2) is faster from the boundary layer to the water column. When currents are weak, the flux of molecules may be diffusion-limited, but after a critical velocity (U_k) is reached, the transfer is no longer limited by diffusion but by the rate of assimilation capacity (i.e. biological or biochemical activity). The mass transfer of molecules also depends on other factors such as the thickness of the periphyton layer on the seagrass leaves, reactions within the periphyton layer and the concentration of molecules in the water adjacent to the leaf-periphyton assemblage.

At the scale of shoots (mm to cm), a feedback mechanism operates as individual shoots are affected by the other shoots and its position within the entire canopy (i.e. edge versus centre of the entire meadow). As water velocity increases, shoots bend which minimises drag, but the forces exerted on individual shoots are more complex when waves are involved as a shoot is exposed to unsteady flows in different directions. This is confirmed by the fact that in wave-swept environments, seagrass leaves become longer as wave exposure increases (de Boer 2007). Flow around shoots results in bending but also pressure gradients on the leeward side of the leaf such that a vertical ascending flow is generated downstream of the shoot. This water then disperses horizontally at the point where the leaves bend over with the flow. Interstitial water is also flushed out at the base of the shoot due to the pressure gradients generated on the sediment surface.

At the whole-canopy level, reduced flows occur within the canopy due to the deflection of the current over the canopy and a loss of momentum within the canopy (van Katwijk et al. 2010). Water speed as a result can be 2 to >10 times slower than outside the meadow. It is this process that allows water and sediment particles to be trapped during low tide; even short seagrass canopies can still reduce water velocity (e.g. *Zostera novazelandica*; Heiss et al. 2000). Vertically, however, water flow intensifies at the height of the sheath or stem as these parts are much less effective at reducing water velocity compared with the leaf component. Canopy flow is nevertheless complex because it is a function of the drag or resistance of the leaves on the water.

Seagrass canopies are overall areas where sediments deposit and carbon accumulates largely due to the reduction in velocity and intensity of turbulence, that is, a reduction in flow strength that leads to a reduction in resuspension within the canopy (de Boer 2007). Although few data (Gacia et al. 2003) exist for empirical measurements of sediment deposition in seagrasses, Duarte et al. (2013a, b) estimate a mean rate of 0.2 ± 0.04 cm year^{-1}. Accumulation may be seasonal, especially during summer when seagrasses are at their maximum density and in winter then resuspension may be greater than accumulation when seagrasses are minimal, although roots and rhizomes may alone be sufficient to stabilise the accumulated deposits (Bos et al. 2007). Epiphytes on seagrass leaves may foster the accumulation of sediment particles by increasing the roughness of the canopy and increasing the thickness of the boundary layer on the leaf surface. However, in highly wave-exposed locations, seagrasses may not accumulate fine sediments due to resuspension. Indeed, in some cases, sediment may be coarser beneath seagrass patches as a result of turbulence generated by the leaves themselves.

4.2   Carbon Sequestration

Rates of carbon sequestration in seagrass meadows ($n = 396$) average 220.7 ± 20.1 g C_{org} m^{-2} year^{-1} (± 1 SE) and a median of 167.4 g C_{org} m^{-2} year^{-1} with values ranging from -2094 to 2124 g C_{org} m^{-2} year^{-1} (Table 4.1). As there

Table 4.1 Estimates of annual carbon sequestration rates (g C_{org} m^{-2} year^{-1}) in seagrass meadows worldwide. Duarte et al. (2010) were the source for most of these data in addition to those listed below

Location[source]	Number of observations	Range	Mean
Florida[a]	117	−272 to 1371	71.0
Spain[b]	61	−323 to 620	781.5
Texas[c]	49	−1282 to 1713	−58.4
France[d]	25	−918 to 2335	284.7
Australia[e]	24	−402 to 629	148.8
Denmark[f]	23	−494 to 813	3.0
Mexico[g]	21	−2094 to 1147	233.1
Chesapeake Bay[h]	20	−337 to 1696	377.0
Alabama[i]	14	−475 to 1462	502.3
Indonesia[j]	12	−1434 to 77	−578.5
Greece[k]	6	119–280	172.8
Japan[l]	6	1.8–10.1	5.0
Philippines[m]	4	−34 to 615	213.2
New England[n]	4	−27 to 41	0.7
North Carolina[o]	3	10–66	33.9
The Bahamas[p]	3	731–972	835.9
Norway[q]	2	−0.5 to 44	21.6
Puerto Rico[r]	2	686–2057	1371.0
Portugal[s]	2	276.5–403.2	339.8
Malta[t]	2	133–249	191.0
India[u]	1		2124.2
South Africa[v]	1		365.8
Wadden Sea[w]	1		33.0
The Netherlands[x]	1		51.1
Padilla Bay, Washington[y]	1		−54.9
Corsica[z]	1		41.3
Mauritania[aa]	1		2068.3

(continued)

Table 4.1 (continued)

Location[source]	Number of observations	Range	Mean
Panama[ab]	1		374.7
Italy[ac]	1		153.3

[a]Sources: Odum (1956), Kenworthy and Thayer (1984), Cebrian (2002), Barrón et al. (2004), Martin et al. (2005), Calleja et al. (2006), Stutes et al. (2007), Herbert and Fourqurean (2008), Yarbro and Carlson (2008), and Long et al. (2015)

[b]Sources: Romero et al. (1994), Mateo et al. (1997), Cebrian et al. (2000), Cebrian (2002), Barrón et al. (2004), Holmer et al. (2004), Gazeau et al. (2005), Vaquer-Sunyer et al. (2012), and Hendriks et al. (2014)

[c]Sources: Odum (1962, 1963) and Ziegler and Benner (1999)

[d]Sources: Frankingnoulle and Bouqueneau (1987), Viaroli et al. (1996), Mateo et al. (1997), Welsh et al. (2000); Desious-Paoli et al. (2001), Ouisse et al. (2014), Cox et al. (2016), Delgard et al. (2016), and Olive et al. (2016)

[e]Sources: Moriarty et al. (1990), Eyre and Ferguson (2002), Adams et al. (2016), Rozaimi et al. (2016), and Serrano et al. (2016)

[f]Source: Risgaard-Petersen and Ottosen (2000)

[g]Source: Reyes and Merino (1991)

[h]Sources: Murray and Wetzel (1987), Caffrey (2004), Lee-Nagel (2007), and Rheuban et al. (2014)

[i]Source: Anton et al. (2009)

[j]Source: Erftemeijer et al. (1993)

[k]Source: Apostolaki et al. (2010)

[l]Source: Apostolaki et al. (2014)

[m]Source: Miyajima et al. (2015)

[n]Source: Gacia et al. (2005)

[o]Sources: Caffrey (2004) and Howarth et al. (2014)

[p]Sources: Kenworthy and Thayer (1984) and Cebrian (2002)

[q]Source: Koch and Madden (2001)

[r]Source: Duarte et al. (2002)

[s]Source: Odum (1959)

[t]Sources: Alexandre et al. (2012) and Bahlmann et al. (2015)

[u]Source: Serrano et al. (2016)

[v]Source: Qasim and Bhattathiri (1971)

[w]Source: Baird and Ulanowicz (1993)

[x]Source: Asmus et al. (2000)

[y]Source: Pellikaan and Nieuhuis (1988)

[z]Source: Caffrey 2004 and Caffrey et al. (2014)

[aa]Source: Champenois and Borges (2012)

[ab]Source: Clavier et al. (2014)

[ac]Source: López-Calderón et al. (2013)

are comparatively few sequestration rates derived from dating using radionuclides (Romero et al. 1994; Mateo et al. 1997; Miyajima et al. 2015; Rozaimi et al. 2016), most of these numbers were derived from metabolic measurements of annual primary production and community respiration to determine the amount of carbon available for storage (Cebrian 2002, Duarte et al. 2010, 2013a, b). Like salt marshes and mangroves, there is no clear relationship with latitude as many of the most luxuriant seagrass meadows are composed of *Posidonia oceanica* in the Mediterranean. The data are skewed towards seagrasses of Florida, Spain and Texas, but there are seagrass beds at nearly all locations that show net heterotrophy (those with negative values in Table 4.1), that is, more loss of carbon via respiration than fixed by the plants. Unlike the data for salt marshes and mangroves, nearly all of the seagrass data were derived from metabolic studies rather than from empirical measurements of actual carbon storage; thus these data do not necessarily account for possible export of 'excess' carbon fixed by the plants nor possible import of carbon from adjacent ecosystems, such as mangroves, salt marshes, coral reefs, rivers or oceanic inputs. Nevertheless, on average, seagrass meadows store carbon although apparently less than salt marshes and mangroves. This conclusion was also reached for tropical Indo-Pacific seagrasses, with an estimated average net sink of 155 g C_{org} m^{-2} $year^{-1}$ (Unsworth et al. 2012).

The sequestration of seagrass carbon is likely to be underestimated as seagrasses export a substantial portion of their primary production, both in particulate and dissolved form. Available evidence indicates that the export of seagrass carbon represents a significant contribution for carbon sequestration in sediments outside seagrass meadows and in the deep sea (Duarte and Krause-Jensen 2017).

The effects of physical disturbance on carbon sequestration capacity of seagrasses has recently been experimentally determined by Dahl et al. (2016). In a series of field experiments testing the impact of shading and simulated grazing, they found that treatments of high-intensity shading and high-intensity clipping to simulate grazing show significantly lower net community production and carbon content in below-ground biomass than in control plots. This latter effect was caused by erosion of the surface sediment due to the removal of above-ground biomass. Their findings indicate that high-intensity disturbances reduce the ability of seagrass meadows to sequester carbon.

Seagrasses, unlike their marsh and mangrove counterparts, can clearly modify seawater pH to the extent that this phenomenon may have some bearing on their ability to withstand ocean acidification. Near a natural volcanic vent off the Italian coast, Apostolaki et al. (2014) found that at high CO_2 levels in close proximity to the vent, seagrasses have high rates of primary productivity but less biomass possibly due to greater grazing, nutrient limitation or poor environmental conditions. A similar result was found in relation to a CO_2 vent in Papua New Guinea (Russell et al. 2013). Thus, seagrass responses to ocean acidification may be complex rather than a simple overall positive or negative reaction. The capacity of seagrasses to modify their ambient pH may have implications for nearby coral reefs as the presence of seagrasses results in a net increase in pH possibly ameliorating the impacts of acidification (Unsworth et al. 2012).

The net increase in pH suggests a positive trend between seagrass productivity and carbonate deposition. Indeed, a study of particulate inorganic carbon (PIC, mostly $CaCO_3$) in seagrasses has shown that PIC stocks in the top 1 m of sediment average 654 Mg PIC ha^{-1} exceeding POC (particulate organic carbon) stocks by a factor of 5 (Mazarrasa et al. 2015). Meadows dominated by *Halodule, Thalassia* or *Cymodocea* support the highest PIC stocks which decrease polewards by 8 Mg PIC ha^{-1} per degree of latitude. Using PIC sediment stocks and estimates of sediment accretion, Mazarrasa et al. (2015) estimated a mean PIC accumulation rate of 126.3 g PIC m^{-2} year^{-1} or roughly one-half of the estimated rate of organic carbon sequestration (Table 4.1). Further, based on the global extent of seagrasses (177,000 to 600,000 km^2), seagrasses globally store between 11 and 39 Pg PIC in the top metre of sediment and accumulate between 22 and 75 Tg PIC year^{-1}. This range of values suggests a significant contribution to coastal carbonate carbon sequestration by seagrasses (Gullström et al. 2018; Howard et al. 2018). High rates of carbonate accumulation imply CO_2 emissions from precipitation, but the POC and PIC stocks between vegetated and un-vegetated sediments demonstrate that seagrass meadows are strong overall CO_2 sinks.

4.3   Carbon Stocks

Published and unpublished measurements of the organic carbon content of living seagrass biomass and underlying soils were compiled recently by Fourqurean et al. (2012a, b) based on data from seagrass meadows across the globe. The results show a wide spread of data of soil organic carbon storage (Fig. 4.1) with most observations being <100 Mg C_{org} ha^{-1} from short (<1 m) cores, but much higher carbon inventories from cores taken to at least 1 m depth. Overall, a median value of 69.3 Mg C_{org} ha^{-1} was derived. Median above- and below-ground biomass were 0.264 and 0.540 Mg C_{org} ha^{-1}, respectively, underscoring that nearly all seagrass organic carbon is stored in soil.

Geographically, it is difficult to discern true trends or patterns in the data owing to the scarcity of data from many parts of the globe. Nevertheless, it is clear that meadows of the Mediterranean seagrass *Posidonia oceanica* have the highest average soil storage (372.4 Mg C_{org} ha^{-1}). The median soil C_{org} stock value is about equal to the average for terrestrial soils, but about one-fourth of the median for salt marsh soils and one-tenth that of the median for mangrove soils.

Using estimates of global seagrass area of between 300,000 and 600,000 km^2 and multiplying by the median soil C_{org} value, we derive a range of global C_{org} values of between 2.1 and 4.2 Pg C_{org} for soils and between 75.5 and 151 Tg C_{org} for biomass. If we assume that the 1 m soil data is the most complete inventory, the soil C_{org} stock rises to between 5.8 and 9.8 Pg C_{org}. As with salt marshes and mangroves, soil C_{org} stocks can be much greater in systems where unconsolidated soils accumulate to depths greater than 1 m, such as in *Posidonia oceanica* meadows where 11 m thick deposits have been found. Of course, meadows growing on coarse carbonates may

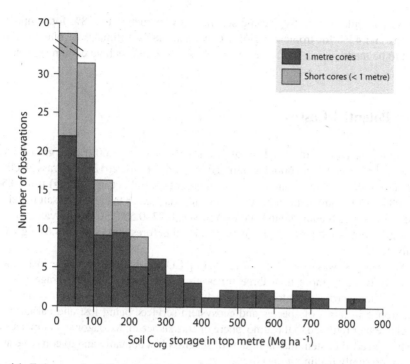

Fig. 4.1 Estimates of soil C_{org} stored in the world's seagrass meadows. Updated from Fourqurean et al. (2012a) using data from Lavery et al. (2013), Campbell et al. (2015), Miyajima et al. (2015), Phang et al. (2015), Alongi et al. (2016), Rozaimi et al. (2016) and Serrano et al. (2016). Bars without shading indicate estimates made based on shallow (<1 m) sediment cores, and black shading indicates estimates based on 1 m length cores

have fairly shallow deposits of less than a metre with correspondingly small C_{org} but large carbonate carbon inventories (Campbell et al. 2015).

Carbon storage in seagrass soils is a reflection of long-term nutrient history. In a comparison of long-term nutrient history versus short-term nutrient enrichment, Armitage and Fourqurean (2016) found that in sites undergoing 17 months of nutrient additions, biomass carbon both above- and below-ground increase but soil carbon content decrease by about 10% in response to phosphorous addition. There is also less than 3% organic carbon in soil when seagrass leaf N:P exceeds a threshold of 75:1 or when below-ground seagrass carbon stock is less than 100 g m^{-2} in the experimental plots and within a naturally occurring long-term gradient of phosphorus availability. Their results show that even under nutrient-limited conditions, seagrass beds have very high potential for carbon storage.

Black carbon may, in some instances, lead to an overestimation of carbon stocks. Chew and Gallagher (2018) found that failure to subtract allochthonous recalcitrant carbon (black carbon) formed outside the ecosystem overvalues the storage of organic carbon. They estimate that current carbon stock estimates are positively biased, particularly for low organic seagrass environments, by 18% for temperate regions and 43% for tropical regions. Obviously, more estimates of black carbon need to be made in order to more accurately assess seagrass blue carbon stocks.

4.4   Potential Losses

Assuming an annual rate of loss of 7% (Waycott et al. 2009) and global area estimates of 300,000 to 600,000 km^2 (Fourqurean et al. 2012), seagrass decline returns to either the atmosphere or to the adjacent coastal ocean (or both) from 0.54 to 1.08 Pg CO_2 equivalents annually. This range is greater than that for salt marshes (0.02–0.24 Pg CO_2 equivalents) and mangroves (0.27–0.59 Pg CO_2 equivalents) and equal to about one-quarter of the average annual deforestation rate of 4.61 Pg CO_2 equivalents.

If all seagrass was destroyed, 7.7 to 15.4 Pg CO_2 equivalents would be lost which is nearly twice to more than three times greater than the annual average rate of deforestation across the globe. Obviously, the loss of seagrass is an ecological catastrophe in terms of species and ecosystem services being lost and carbon that is being either returned to the atmosphere or coastal ocean. Management emphasis is urgently needed to stem the high rates of seagrass lost annually and to conserve and restore presently declining meadows.

References

Adams M, Ferguson A, Maxwell PS, O'Brian KR (2016) Light history-dependent respiration explains the hysteresis in the daily ecosystem metabolism of seagrasses. Hydrobiologia 766:75–88

Alexandre A, Silva J, Buapet P, Björk M, Santos R (2012) Effects of CO_2 enrichment on photosynthesis, growth and nitrogen metabolism of the seagrass *Zostera noltii*. Ecol Evol 2:2625–2635

Alongi DM, Murdiyarso D, Fourqurean JW, Kauffman JB, Hutahaean A, Crooks S, Lovelock CE, Howard J, Herr D, Fortes M, Pidgeon E, Wagey T (2016) Indonesia's blue carbon: a globally significant and vulnerable sink for seagrass and mangrove carbon. Wetl Ecol Manage 24:3–13

Anton A, Cebrian J, Duarte CM, Heck J, Kenneth L, Goff J (2009) Low impact of Hurricane Katrina on seagrass community structure and functioning in the Northern Gulf of Mexico. Bull Mar Sci 85:45–59

Apostolaki ET, Holmer M, Marbá N, Karakassis I (2010) Metabolic imbalance in coastal vegetated (*Posidonia oceanica*) and unvegetated benthic ecosystems. Ecosystems 13:459–471

Apostolaki ET, Vizzini S, Hendriks IE, Olsen YS (2014) Seagrass ecosystem response to long-term high CO_2 in a Mediterranean volcanic vent. Mar Environ Res 99:9–15

Armitage AR, Fourqurean JW (2016) Carbon storage in seagrass soils: long-term nutrient history exceeds the effects of near-term nutrient enrichment. Biogeosciences 13:313–321

Asmus RM, Sprung M, Asmus H (2000) Nutrient fluxes in intertidal communities of a South European lagoon (Ria Formosa): similarities and differences with a northern Wadden Sea bay (Sylt-Romo Bay). Hydrobiologia 436:217–235

Bahlmann E, Weinberg I, Lavrič JV, Eckhardt T, Michaelis W, Santos R, Seifert R (2015) Tidal controls on trace gas dynamics in a seagrass meadow of the Rio Formosa lagoon (southern Portugal). Biogeosciences 12:1683–1696

Baird D, Ulanowicz RE (1993) Comparative study on the trophic structure, cycling and ecosystem properties of four tidal estuaries. Mar Ecol. Prog Ser 99:221–237

Barrón C, Marbá N, Terrados J, Kennedy H, Duarte CM (2004) Community metabolism and carbon budget along a gradient of seagrass (Cymodocea nodosa) colonization. Limnol Oceangr 49:1642–1651

Bos AR, Bouma TJ, de kort GLJ, van Katwijk MM (2007) Ecosystem engineering by annual intertidal seagrass beds: sediment accretion and modification. Estuar Coast Shelf Sci 74:344–348

Caffrey JM (2004) Factors controlling net ecosystem metabolism in U.S. estuaries. Estuaries 27:90–101

Caffrey JM, Murrell MC, Amacker KS, Harper JW, Philipps S, Woodrey MS (2014) Seasonal and inter-annual patterns in primary production, respiration, and net ecosystem metabolism in three estuaries in the Northeast Gulf of Mexico. Estuar Coast 37:222–241

Calleja ML, Barrón C, Hale JA, Frazer TK, Duarte CM (2006) Light regulation of benthic sulfate reduction rates mediated by seagrass (Thalassia testudinum) metabolism. Estuar Coast 29:1255–1264

Campbell JE, Lacey EA, Decker RA, Crooks S, Fourqurean JW (2015) Carbon storage in seagrass beds of Abu Dhabi, United Arab Emirates. Estuar Coast 38:242–251

Cebrian J (2002) Variability and control of carbon consumption, export, and accumulation in marine communities. Limnol Oceanogr 47:11–22

Cebrian J, Pedersen MF, Kroeger KD, Valiela I (2000) Fate of production of the seagrass Cymodocea nodosa in different stages of meadow formation. Mar Ecol Prog Ser 204:119–130

Champenois W, Borges A (2012) Seasonal and inter-annual variations of community metabolism rates of a Posidonia oceanica meadow. Limnol Oceangr 57:347–361

Chew ST, Gallagher JB (2018) Accounting for black carbon lowers estimates of blue carbon storage services. Sci Rept 8:2553–2560

Clavier J, Chauvard L, Amice E, Lazure P, van der Geest M, Labrosse P, Diagne A, Carlier A, Chauvard S (2014) Benthic metabolism in a shallow coastal ecosystem of the Banc d'Arguin. Mar Ecol Prog Ser 501:11–23

Cox TE, Gazeau F, Allioune S, Hendriks IE, Makacek P, Le Fur A, Gattuso JP (2016) Effects of in situ CO_2 enrichment on structural characteristics, photosynthesis, and growth of the Mediterranean seagrass Posidonia oceanica. Biogeosciences 13:2179–2194

Cullen-Unsworth LC, Nordlund LM, Paddock J, Baker S, McKenzie LJ, Unsworth RKF (2014) Seagrass meadows globally as a coupled social-ecological system: implications for human wellbeing. Mar Pollut Bull 83:387–397

Dahl M, Deyanova D, Lyimo LD, Näslund J, Samuelsson GS, Mtolera MSP, Björk M, Gullström M (2016) Effects of shading and simulating grazing on carbon sequestration in a tropical seagrass meadow. J Ecol 104:654–664

de Boer WF (2007) Seagrass-sediment interactions, positive feedbacks and critical thresholds for occurrence: a review. Hydrobiologia 591:5–24

Delgard ML, Deflandre B, Bernard G, Richard M, Kockoni E, Charbonnier C, Cesbrom F, Metzger E, Gremare A, Anschultz P (2016) Benthic oxygen exchange over a heterogeneous Zostera noltei meadow in a temperate coastal ecosystem. Mar Ecol Prog Ser 543:55–71

Desious-Paoli JM, Auby I, Dagault F (2001) Factors influencing primary production of seagrass beds (Zostera noltii Hornem.) in the Thau lagoon (French Mediterranean coast). J Exp Mar Biol Ecol 259:63–84

Duarte CM, Krause-Jensen D (2017) Export of seagrass meadows contributes to marine carbon sequestration. Front Mar Sci 4:13. https://doi.org/10.3389/fmars.2017.00013

Duarte CM, Martinez R, Barrón C (2002) Biomass, production and rhizome growth near the northern limit of seagrass (Zostera marina) distribution. Aq Bot 72:183–189

Duarte CM, Middelburg JJ, Caraco N (2005) Major role of marine vegetation on the oceanic carbon cycle. Biogeosciences 2:1–8

Duarte CM, Marbä N, Gacia E, Fourqurean JW, Beggins J, Barrón C, Apostolaki ET (2010) Seagrass community metabolism: assessing the carbon sink capacity of seagrass meadows. Glob Biogeochem Cycles 24 https://doi.org/10.1029/2010GB003793

Duarte CM, Kennedy H, Marbá N, Hendriks I (2011) Assessing the capacity of seagrass meadows for carbon burial: current limitations and future strategies. Ocean Coast Manage 83:32–38

Duarte CM, Losada IJ, Hendriks IE, Mazarrasa I, Marbá N (2013a) The role of coastal plant communities for climate change mitigation and adaptation. Nat Climate Change 3:961–968

Duarte CM, Sintes T, Marbá N (2013b) Assessing the CO_2 capture potential of seagrass restoration projects. J Appl Ecol 50:1341–1349

Erftemeijer PLA, Osinga R, Mars AE (1993) Primary production of seagrass beds in South Sulawesi (Indonesia): a comparison of habitats, methods and species. Aq Bot 46:67–90

Eyre BD, Ferguson JP (2002) Comparison of carbon production and decomposition, benthic nutrient fluxes and denitrification in seagrass, phytoplankton, benthic microalgae- and macroalgae-dominated warm temperate Australian lagoons. Mar Ecol Prog Ser 229:43–59

Fourqurean JW, Duarte CM, Kennedy H, Marbá MH, Mateo MA, Apostolaki ET, Kendrick GA, Krause-Jensen D, McGlathery KJ, Serano O (2012a) Seagrass ecosystems as a globally significant carbon stock. Nature Geosci 5:505–509

Fourqurean JW, Kendrick GA, Collins LS, Chambers RM, Vanderklift MA (2012b) Carbon, nitrogen and phosphorus storage in subtropical seagrass meadows: examples from Florida Bay and Shark Bay. Mar Freshw Res 63:967–983

Frankingnoulle M, Bouquegneau JM (1987) Seasonal variation of the diel carbon budget of a marine macrophyte ecosystem. Mar Ecol Prog Ser 38:197–199

Gacia E, Duarte CM, Marbá N, Terrados J, Kennedy H, Fortes MD, Tri NH (2003) Sediment deposition and production in SE-Asia seagrass meadows. Estuar Coast Shelf Sci 56:909–919

Gacia E, Kennedy H, Duarte CM, Terrados J, Marbá N, Papadimitriou S, Fortes M (2005) Light-dependence of the metabolic balance of a highly productive Philippine seagrass community. J Exp Mar Biol Ecol 316:55–67

Gazeau F, Duarte CM, Gattuso JP, Barrón C, Navarro N, Ruiz S, Borges AV (2005) Whole-system metabolism and CO_2 fluxes in a Mediterranean Bay dominated by seagrass beds (Palma Bay, NW Mediterranean). Biogeosciences 2:43–60

Gullström M, Lyimo LD, Dahl M, Samuelsson GS, Eggertsen M, Anderberg E, Rasmusson LM, Linderholm HW, Knudby A, Banderia S, Nordlund LM, Björk M (2018) Blue carbon storage in tropical seagrass meadows relates to carbonate stock dynamics, plant-sediment processes, and landscape context: insights from the Western Indian Ocean. Ecosystems 21:551–566

Heiss WM, Smith AM, Probert PK (2000) Influence of the small intertidal seagrass *Zostera novazelandica* on linear water flow and sediment texture. New Zealand J Mar Freshw Res 34:689–694

Hendriks IE, Olsen YS, Ramajo L, Basso L, Steckbauer A, Moore TS, Howard J, Duarte CM (2014) Photosynthetic activity buffers ocean acidification in seagrass meadows. Biogeosciences 11:333–346

Herbert DA, Fourqurean JW (2008) Ecosystem structure and function still altered two decades after short-term fertilization of a seagrass meadow. Ecosystems 11:688–700

Holmer M (2009) Productivity and biogeochemical cycling in seagrass ecosystems. In: Perillo GME, Wolanski E, Cahoon DR, Brinson MM (eds) Coastal wetlands: an integrated ecosystem approach. Elsevier, Amsterdam, pp 377–401

Holmer M, Duarte CM, Boschker HTS, Barrón C (2004) Carbon cycling and bacterial carbon sources in pristine and impacted Mediterranean seagrass sediments. Aq Micro Ecol 36:227–237

Howard JL, Creed JC, Agular MVP, Fourqurean JW (2018) CO_2 released by carbonate sediment production in some coastal areas may offset the benefits of seagrass "Blue Carbon" storage. Limnol Oceanogr 63:160–172

Howarth RW, Hayn M, Marino RM, Ganju N, Foreman K, McGlathery K, Giblin AE, Berg P, Walker JD (2014) Metabolism of a nitrogen-enriched coastal lagoon during the summertime. Biogeochemistry 118:1–20

Jackson EL, Rowden AA, Attrill MJ, Bossey S, Jones M (2001) The importance of seagrass beds as a habitat for fishery species. Oceanogr Mar Biol Annu Rev 39:269–304

Kenworthy J, Thayer G (1984) Production and decomposition of the roots and rhizomes of seagrasses *Zostera marina* and *Thalassia testudinum* in temperate and subtropical marine ecosystems. Bull Mar Sci 35:364–379

Koch MS, Madden CJ (2001) Patterns of primary productivity and nutrient availability in a Bahamas lagoon with fringing mangroves. Mar Ecol Prog Ser 219:109–119

Koch EW, Ackerman JD, Verduin J, van Keulen M (2006) Fluid dynamics in seagrass ecology – from molecules to ecosystems. In: Karkum AWD (ed) Seagrasses: biology, ecology and conservation. Springer, Dordrecht, pp 193–225

Lavery PS, Mateo MA, Serrano O, Rozaimi M (2013) Variability in the carbon storage of seagrass habitats and its implications for global estimates of blue carbon ecosystem service. PLoS One 8: e73748

Lee-Nagel JL (2007) Plant-sediment interactions and biogeochemical cycling for seagrass communities in Chesapeake and Florida Bays. PhD thesis, Univer of Maryland, USA

Long MH, Berg P, Falter JL (2015) Seagrass metabolism across a productivity gradient using eddy covariance, Eulerian control volume, and biomass addition techniques. J Geophys Res (Oceans) 120:3624–3639

López-Caldefon C, Guzman HM, Jacome GE, Barnes PAG (2013) Decadal increase in seagrass biomass and temperature at the CARICOMP site in Bocas del Toro, Panama. Rev Biol Trop 61:1815–1826

Martin S, Clavier J, Guarini JM, Chauvaud L, Hily C, Grall J, Thouzeau G, Jean F, Richard J (2005) Comparison of *Zostera marina* and maerl community metabolism. Aq Bot 83:161–174

Mateo MA, Romero J, Pérez M, Littler MM, Littler DS (1997) Dynamics of millenary organic deposits resulting from the growth of the Mediterranean seagrass *Posidonia oceanica*. Estuar Coast Shelf Sci 44:103–110

Mateo MA, Cebrian J, Dunton KH, Mutchler T (2006) Carbon flux in seagrass ecosystems. In: Larkum AWD (ed) Seagrasses: biology, ecology and conservation. Springer, Dordrecht, pp 159–192

Mazarrasa I, Marbá N, Lovelock CE, Serrano O, Lavery PS, Fourqurean JW, Kennedy H, Mateo MA, Krause-Jensen D, Steven ADL, Duarte CM (2015) Seagrass meadows as a globally significant carbonate reservoir. Biogeosciences 12:4993–5003

Miyajima T, Hori M, Hamaguchi M, Shimabukuro H, Adachi H, Yamano H, Nakaoka M (2015) Geographic variability in organic carbon stock and accumulation rate in sediments of East and Southeast Asian seagrass meadows. Global Biogeochem Cycles 29:397–415

Moriarty DJW, Roberts DG, Pollard PC (1990) Primary and bacterial productivity of tropical seagrass communities in the Gulf of Carpentaria, Australia. Mar Ecol Prog Ser 61:145–157

Murray L, Wetzel RL (1987) Oxygen production and consumption associated with the major autotrophic components in two temperate seagrass communities. Mar Ecol Prog Ser 38:231–239

Odum HT (1956) Primary production in flowing waters. Limnol Oceanogr 1:102–117

Odum HT (1959) Measurements of productivity of turtle grass flats, reefs and the Bahia Fosforescente of southern Puerto Rico. Inst Mar Sci Univ Texas 6:159–170

Odum HT (1962) Further studies on reaeration and metabolism of Texas Bays 1958-1960. Inst Mar Sci Univ Texas 8:23–55

Odum HT (1963) Productivity measurements in Texas turtle grass and the effects of dredging an intracoastal channel. Inst Mar Sci Univ Texas 6:48–58

Olive I, Silva J, Costa MM, Santos R (2016) Estimating seagrass community metabolism using benthic chambers: the effects of incubation time. Estuar Coasts 39:138–144

Ouisse V, Migné A, Davoult D (2014) Comparative study of methodologies to measure in situ the intertidal benthic community metabolism during immersion. Estuar Coast Shelf Sci 136:19–25

Pellikaan G, Nienhuis P (1988) Nutrient uptake and release during growth and decomposition of eelgrass, *Zostera marina* L., and its effect on the nutrient dynamics of Lake Grevelingen. Aq Bot 30:189–214

Phang VXH, Chou LM, Friess DA (2015) Ecosystem carbon stocks across a tropical intertidal habitats mosaic of mangrove forest, seagrass meadow, mudflat and sandbar. Earth Surf Process Landf 40:1387–1400

Qasim SZ, Bhattathiri PMA (1971) Primary production of a seagrass bed on Kavaratti Atoll (Laccadives). Hydrobiologia 38:29–38

Reyes E, Merino M (1991) Diel dissolved oxygen dynamics and eutrophication in a shallow well-mixed tropical lagoon (Cancun, Mexico). Estuaries 14:372–381

Rheuban JE, Berg˙P, McGlathery KJ (2014) Multiple timescale processes drive ecosystem metabolism in eelgrass (*Zostera marina*) meadows. Mar Ecol Prog Ser 507:1–13

Risgaard-Petersen N, Ottosen LDM (2000) Nitrogen cycling in two temperate *Zostera marina* beds: seasonal variation. Mar Ecol Prog Ser 198:93–107

Romero J, Pérez M, Mateo MA, Sala A (1994) The belowground organs of the Mediterranean seagrass *Posidonia oceanica* as a biogeochemical sink. Aq Bot 47:13–19

Rozaimi M, Lavery PS, Serrano O, Kyrwood D (2016) Long-term carbon storage and its recent loss in an estuarine *Posidonia australis* meadow (Albany, Western Australia). Estuar Coast Shelf Sci 171:58–65

Russell BD, Connell SD, Uthicke S, Muehllehner N, Fabricius KE, Hall-Spencer JM (2013) Future seagrass beds: can increased productivity lead to increased carbon storage? Mar Pollut Bull 73:463–469

Samper-Villarreal J, Lovelock CE, Saunders MI, Roelfsema C, Mumby PJ (2016) Organic carbon on seagrass sediments is influenced by seagrass canopy complexity, turbidity, wave height, and water depth. Limnol Oceanogr 61:938–952

Serrano O, Lavery PS, López-Merino L, Ballesteros E, Mateo MA (2016) Location and associated carbon storage of erosional escarpments of seagrass *Posidonia* mats. Front Mar Sci 3:42

Short FT, Polidoro B, livingstone SR, Carpenter KE, Bandeira S, Bujang JS, Calumpong HP, Carruthers TJB, Coles RG, Dennison WC, Erftemeijer PLA, Fortes MD, Freeman AS, Jagtap TG, Kamal AHM, Kendrick GA, Judson Kenworthy W, La Nafie TA, Nasution IM, Orth RJ, Prathep TG, Sanciangco JC, Tussenbroeck BV, Vergara SG, Waycott M, Zieman JC (2011) Extinction risk assessment of the world's seagrass species. Biol Conserv 144:1961–1971

Stutes J, Cebrian J, Stutes AL, Hunter A, Corcoran AA (2007) Benthic metabolism across a gradient of anthropogenic impact in three shallow coastal lagoons in NW Florida. Mar Ecol Prog Ser 348:55–70

Unsworth RKF, DeLeon PS, Garrard SL, Jompa J, Smith DJ, Bell JJ (2008) High connectivity of Indo-Pacific seagrass fish assemblages with mangrove and coral reef habitats. Mar Ecol Prog Ser 353:213–224

Unsworth RKF, Collier CJ, Henderson GM, McKenzie LJ (2012) Tropical seagrass meadows modify seawater carbon chemistry: implications for coral reefs impacts by ocean acidification. Environ Res Lett 7. https://doi.org/10.1088/1748-9326/7/2/024026

Unsworth RKF, van Keulen M, Coles RG (2014) Seagrass meadows in a globally changing environment. Mar Pollut Bull 83:383–386

van Katwijk MM, Bos AR, Hermus DCR, Suykerbuyk W (2010) Sediment modification by seagrass beds: muddification and sandification induced by plant cover and environmental conditions. Estuar Coast Shelf Sci 89:175–181

Vaquer-Sunyer R, Duarte CM, Jordá G, Ruiz-Halpern S (2012) Temperature dependence of oxygen dynamics and community metabolism in a shallow Mediterranean macroalgal meadow (*Caulerpa prolifera*). Estuar Coast 35:1182–1192

Viaroli P, Bartoli M, Bondavalli C, Christian RR, Giordani G, Naldo M (1996) Macrophyte communities and their impact on benthic fluxes of oxygen, sulphide and nutrients in shallow eutrophic environments. Hydrobiologia 329:105–119

Waycott M, Duarte CM, Carruthers TJB, Orth RJ, Dennison WC, Olyarnik S, Calladine A, Fourqurean JW, Heck KL Jr, Hughes AR, Kendrick GA, Kenworthy WJ, Short FT, Williams SL (2009) Accelerating loss of seagrasses across the globe threatens coastal ecosystems. Proc Nat Acad Sci USA 106:12377–12381

Welsh DT, Bartoli M, Nizzoli D, Castaldelli G, Riou SA, Viaroli P (2000) Denitrification, nitrogen fixation, community primary productivity and inorganic-N and oxygen fluxes in an intertidal *Zostera noltii* meadow. Mar Ecol Prog Ser 208:65–77

Yarbro LA, Carlson PRJ (2008) Community oxygen and nutrient fluxes in seagrass beds of Florida Bay, USA. Estuar Coast 31:877–897

Ziegler S, Benner R (1999) Nutrient cycling in the water column of a subtropical seagrass meadow. Mar Ecol Prog Ser 188:51–62

Chapter 5
Kelp Forests

Kelp forests are comprised primarily of brown algae in the order Laminariales and dominate shallow rocky coasts of the world's cold-water regions. Kelp beds are phyletically diverse, structurally complex and highly productive ecosystems containing a wide variety of marine mammals, sea urchins, fishes, crabs, molluscs, other algae and epibiota. Their global distribution is physiologically constrained by light at high latitudes and by nutrients, warm temperatures and other macrophytes at low latitudes.

Kelps belong to three morphological groups or guilds defined by canopy height: canopy, stipitate and prostrate forms (Steneck et al. 2002). Canopy kelps are the largest type, producing floating canopies which can grow up to 45 m long. *Macrocystis* spp. is the chief genus dominating kelp forests along the west coasts of North and South America and at various locations such as South Africa, southern Australia, New Zealand and several subantarctic islands. Smaller canopy kelps (about 10 m height), such as *Nereocystis luetkeana*, range from central California to Alaska, while its Southern Hemisphere counterparts, *Ecklonia maxima* and *Alaria fistulosa*, occur in South Africa and Alaska and the Pacific coast of Asia, respectively. Stipitate kelps are held above the seabed by rigid stipes and include some species of *Laminaria* in Europe and the Pacific north and northwest, *Ecklonia* in southern Australia and New Zealand and *Lessonia* in Chile. These forests can grow from 5 to 10 m in height. Prostrate kelps are the smallest and cover the seabed with their fronds. This group includes several species of *Laminaria* which range from the Gulf of Maine to Greenland and from Iceland to the high Arctic of Norway and south to the northwesternmost corner of Africa.

Kelp forests profoundly alter local environmental conditions by dampening waves which influence water flow and associated processes of coastal erosion, sedimentation, recruitment and benthic productivity; they also reduce light, creating an understorey favourable for many species that prefer low light conditions. Kelp architecture provides a habitat, nursery ground and food for a wide variety of organisms, both pelagic and benthic. Kelps are highly productive and are a significant source of nutrition via macroalgal detritus; only rarely do herbivores consume

© The Author(s) 2018
D. M. Alongi, *Blue Carbon*, SpringerBriefs in Climate Studies,
https://doi.org/10.1007/978-3-319-91698-9_5

more than 10% of living kelp biomass. Kelp detritus litters the seabed, becoming food for detritivores and microbes, thus concentrating and magnifying secondary production and supporting complex food webs in the coastal zone.

Three interacting processes control the development of kelp forests: recruitment, growth and competition. Recruitment is often seasonal and influenced by environmental conditions at the time of settlement of zygotes. Growth of kelps depends on interactions among nutrient availability, temperature and light. Kelps dominate cold-water regimes but can become physiologically stressed at high temperatures, especially during periods of low nutrient availability. Kelps have a relatively low photosynthetic to biomass ratio which constrains them to relatively shallow, well-illuminated zones; kelps free of herbivores and other forms of disturbance decline rapidly in frond size and density.

Kelps are mostly threatened by herbivory, usually from sea urchins (Steneck et al. 2002). Widespread kelp destruction occurs when overfishing and the loss of apex predators triggers herbivore population increases. Urchin-induced deforestation has been increasing over the past few decades. However, continued fishing down the food web has resulted in shifting harvesting targets from apex predators to their invertebrate prey including kelp-grazing herbivores. Thus, some areas have seen a return of kelp forests. Overfishing appears to be the greatest manageable threat to kelp forests.

5.1 Do Kelp Forests Sequester and Store Blue Carbon?

Kelp forests are highly productive ecosystems that produce large amounts of fixed carbon. Kelps produce large amounts of detritus through incremental blade erosion, fragmentation of blades and dislodgement of whole fronds and thalli. Rates of primary productivity of kelps range from 9 to 5622 g C m^{-2} year^{-1} with average and median values of 1057 and 664 g C m^{-2} year^{-1}, respectively (Krumhansl and Scheibling 2012). Reed and Brzezinski (2009) estimate a global kelp production of 15 Tg C year^{-1} and that if deep tropical areas are included, then global kelp production approaches 39 Tg C year^{-1}. They further estimate a global standing kelp crop of from 7.5 to 20 Tg C.

The estimated global average rate of detrital production by kelps is 706 g C m^{-2} year^{-1}, accounting for 82% of annual kelp productivity. Rates of detrital production range from 8 to 2657 g C m^{-2} year^{-1}, from blade erosion and fragmentation, and from 22 to 839 g C m^{-2} year^{-1} for loss of fronds and thalli. This production is second only to detrital production in salt marshes and roughly equivalent to production in seagrass beds (Krumhansl and Scheibling 2012). Furthermore, based on nitrogen content and C/N ratio, kelp detritus is more nutritious than salt marsh and mangrove detritus and is more readily consumed and assimilated by detritivores.

Detritus production is regulated by current and wave hydrodynamics and is highest during severe storms and as a result of blade weakening through damage by grazers and encrusting epibionts. Detritus settles within kelp beds and is also exported to adjacent as well as distant habitats, such as sandy beaches and the deep sea. Kelp detritus provides a significant food subsidy and enhances secondary production in kelp beds and in distal communities that can be kilometres from the source.

Detritus from plants such as marine macroalgae do contain significant quantities of carbon resistant to microbial decay (refractory carbon), so although they may themselves not accumulate organic or carbonate carbon, their detritus will be valuable in other adjacent habitats for long-term carbon storage (Trevathan-Tackett et al. 2015).

Kelp beds have been undervalued as prime sites of carbon storage in living biomass (Smale et al. 2016). Off the UK coast, Smale et al. (2016) found that standing stocks of kelp range from 251 to 1820 g C m^{-2} with an average carbon storage of 721 ± 140 g C m^{-2} which is greater than historical estimates. Kelp forests generally do not develop their own soft organic-rich sediments, so they have limited capacity to act as long-term carbon sinks in the traditional sense. However, they may instead act as carbon donors to adjacent benthic ecosystems where kelp detritus accumulates. Hill et al. (2015) recently estimated that 109.9 to 274.7 Tg C is stored in temperate Australian macroalgae (mostly kelps) with an additional estimate of 23.2 Tg C derived for tropical and subtropical regions globally. They point out that the vast bulk of carbon storage in kelp forests is in living above-ground biomass, but there is opportunity for calcifying macroalgae, such as *Halimeda*, to store large quantities of carbonate carbon in macroalgal beds in subtropical and tropical areas. *Halimeda* bioherms have been estimated to contribute 400 Tg C year^{-1}, and rhodolith-forming species can produce from 60 to 1000 g CaCO$_3$ m^{-2} year^{-1}. However, the amounts of carbonate carbon globally are difficult to quantify owing to a lack of data.

The data collated by Krause-Jensen and Duarte (2016) suggest that macroalgae such as kelps can be a significant source of carbon to the deep ocean. Their analysis indicates that as much as 173 TgC year^{-1} (range, 61–268 TgC year^{-1}) of macroalgal carbon can be sequestered globally with about 90% of this sequestration occurring through export to the deep sea and the rest through burial in coastal sediments.

Carbonate carbon has not generally been considered in estimates of blue carbon because the calcification process results in the release of 1 CO$_2$ molecule for every molecule of CaCO$_3$ formed:

$$2HCO_3^- + Ca^{2+} \rightarrow CaCO_3 + CO_2 + H_2O \qquad (5.1)$$

However, the carbon dioxide produced during this process can be rapidly utilised in photosynthesis; even if not, there is still the net conversion of one HCO$_3^-$ molecule into one calcium carbonate molecule, that is, calcification still results in the draw-down of carbon in the minerals aragonite or calcite. Hill et al. (2015) argue that calcification should be considered in any blue carbon calculations in future as, if included, this process would constitute a further sink of 0.34 Pg C year^{-1} globally.

5.2 Kelp Carbon and Climate Change

Macroalgae such as kelp may respond positively to global climate change, including the problem of ocean acidification which may benefit macroalgae that are able to capitalise on increased inorganic carbon availability for photosynthesis (Koch et al. 2013; Celis-Piá et al. 2015) although not all species may show such effects (Fernández et al. 2015). In laboratory experiments, the giant kelp *Macrocystis pyrifera* did not show a change in photosynthetic or growth rates under conditions of elevated CO_2 supply and lower pH (Fernández et al. 2015). This result was explained by the greater use of HCO_3^- compared to CO_2 to support photosynthesis. Along a natural gradient of CO_2 concentrations in proximity to a submarine CO_2 seep off Italy, Celis-Piá et al. (2015) found that other environmental conditions such as light and nutrient levels play a key role in the response of the non-calcifying macroalga *Cystoseira compressa* and the calcifying *Padina pavonica* to ocean acidification. Using a suite of biochemical assays, they found that both algae benefited from elevated CO_2 levels, although their responses varied depending on light and nutrient availability. In *C. compressa*, elevated CO_2 resulted in higher carbon content and antioxidant activity in shaded conditions with and without nutrient enrichment, whereas *P. pavonica* also showed high carbon content, higher photosynthetic efficiency and higher quantum yield in elevated CO_2 treatments, but had higher concentrations of phenolic compounds in nutrient-enriched, fully lit conditions and more antioxidants in shaded nutrient-enriched conditions. Thus, it appears that the responses of kelps and other macroalgae to ocean acidification are species-specific.

In an acidifying marine realm, it may be possible to take advantage of the responses of some macroalgae. Chung et al. (2011) predicted that CO_2 acquisition by marine macroalgae can represent a considerable sink for anthropogenic CO_2 emissions and that harvesting and appropriate use of macroalgal primary production could play a greater role in carbon sequestration than previously believed. About 7.5 to 8 million tonnes wet weight of seaweeds are harvested annually from wild and cultivated sources. Of course, this harvested biomass would have to be used as a means of storage such as in pulp rather than in reuse and return to the atmosphere. These seaweeds can also be used for mitigation and adaptation against global warming as discussed in the next chapter.

References

Celis-Piá PSM, Hall-Spencer JM, Horta PA, Milazzo M, Korbee N, Cornwall CE, Figueroa FL (2015) Macroalgal responses to ocean acidification depends on nutrient and light levels. Front Mar Sci 2:article 26

Chung IK, Beardall J, Mehta S, Sahoo D, Stojkovic S (2011) Using marine macroalgae for carbon sequestration: a critical appraisal. J Appl Phycol 23:877–886

Fernández PA, Roleda MY, Hurd CL (2015) Effects of ocean acidification on the photosynthetic performance, carbonic anhydrase activity and growth of the giant kelp *Macrocystis pyrifera*. Photosynth Res. https://doi.org/10.1007/s11120-015-0138-5

Hill R, Bellgrove A, Macreadie PI, Petrou K, Beardall J, Steven A, Ralph PJ (2015) Can macroalgae contribute to blue carbon? An Australian perspective. Limnol Oceanogr 60:1689–1706

Koch M, Bowes G, Ross C, Zhang XH (2013) Climate change and ocean acidification effects on seagrasses and marine macroalgae. Glob Change Biol 19:103–132

Krause-Jensen D, Duarte CM (2016) Substantial role of macroalgae in marine carbon sequestration. Nature Geosci 9:737–742

Krumhansl KA, Scheibling RE (2012) Production and fate of kelp detritus. Mar Ecol Prog Ser 467:281–302

Reed DC, Brzezinski MA (2009) Kelp forests. In: Laffoley D d'A, Grimsditch G (eds) The management of natural coastal carbon sinks. IUCN, Gland

Smale DA, Burrows MT, Evans AJ, King N, Sayer MDJ, Yunnie ALE, Moore PJ (2016) Linking environmental variables with regional-scale variability in ecological structure and standing stock of carbon within UK kelp forests. Mar Ecol Prog Ser 542:79–95

Steneck RS, Graham MH, Bourque BJ, Corbett D, Erlandson JM, Estes JA, Tegner MJ (2002) Kelp forest ecosystems: biodiversity, stability, resilience and future. Environ Conserv 29:436–459

Trevathan-Tackett SM, Kelleway J, Macreadie PI, Beardall J, Ralph P, Bellgrove A (2015) Comparison of marine macrophytes for their contributions to blue carbon sequestration. Ecology 96:3043–3057

Chapter 6
The Blue Economy: Mitigation and Adaptation

While the science of developing an adequate database for carbon capture and storage in coastal ecosystems has progressed, the application of this data for actual blue carbon projects and the practical running of a blue carbon project for mitigation and adaptation have also progressed, although not without problems; the truth is that most projects have had low success rates (see Sect. 6.2.1). The scientific literature is still divorced from the management literature reflecting the fact that under most circumstances, research and management discussions are held separately. Thomas (2014) noted in an analysis of the blue carbon literature that scientific concepts of mutual relevance cluster together but that user-defined concepts of business, enterprise, finance, funding and costs tend to appear as outliers, with only a single thread linking them to the concept of blue carbon. In other words, the scientific literature of blue carbon is distinct from the management and economic literature. It seems that science and management of blue carbon are nearly mutually exclusive, reflecting the fact that scientists and managers (and business people) are not interacting with regard to blue carbon. Scientists are viewing blue carbon as a science problem and leaving it to managers to apply the science.

These problems have not stopped the push to incorporate blue carbon into what is now called the 'blue economy'. Spaulding (2016) summarises what is driving the new blue economy: a push to adapt the UN Sustainable Development Goals (**SDG**) for the global ocean, especially Goal 14 'Conserve and sustainability use: the oceans, seas and marine resources for sustainable development'. This goal reflects an upgrade on the traditional ocean economy of offshore oil and gas, recreation and commercial fishing, aquaculture, shipping, coastal tourism and telecommunications into renewable energies, remediation/restoration, seabed mining and blue biotechnologies. The new blue economy is thus an upgrade from destructive extraction-focused businesses to sustainable, clean technologies, including blue carbon. The new blue economy is to promote economic benefits of 'good for the ocean' industries and activities while ensuring truly sustainable development. The problem is how to classify these different industries under one umbrella to ensure a level of stewardship, good environmental and social practices and the use of the

© The Author(s) 2018
D. M. Alongi, *Blue Carbon*, SpringerBriefs in Climate Studies,
https://doi.org/10.1007/978-3-319-91698-9_6

precautionary principle in industry to minimise the chance of unsustainable development or industrial accidents or perverse outcomes. Again, the problem remains how to incorporate science into the new blue economy given the reluctance of scientists and managers to cooperate.

This problem is one of several that we will explore in this chapter. Needless to say, it is not uncommon in this author's opinion for scientists and managers to speak different languages and to operate separately. Rare is the environmental problem where scientists and managers work closely together to foster the best outcome. Such linkage is and will remain crucial for successful blue carbon projects.

6.1   Ecological Economics

The 'tragedy of the commons' principle stresses that in a situation within a shared resource system, individual users acting independently (according to their own self-interest) will behave contrary to the common good of all users by depleting that resource through collective action. Wilkinson and Salvat (2012) have asserted that this concept applies to coral reefs, mangrove forests and seagrass beds in the tropics, accounting from much of the degradation of coastal resources. Despite scientific advances in our knowledge of such ecosystems and considerable conservation and management effort, they continue to decline. In the tropics, much of this decline in coastal resources is due to increasing exploitation driven by poverty and progress; in the rest of the world, pollution and so-called economic progress have resulted in a concomitant decline in salt marshes and kelp forests. Thus, the global decline in coastal resources has continued unabated making mitigation and adaptation projects and education a higher priority more than ever. Wilkinson and Salvat (2012) concluded that the solution to the problem will be implementing exceedingly difficult and controversial moral decisions.

With blue carbon, such decisions will need to be made at the national level, but the reality is that humans preserve best what is most financially and culturally valuable to them. The concept of a blue carbon project is economically viable; the high cost/benefit ratios for loss versus conservation of coastal ecosystems are high, with economic damage resulting from conversion currently amounting to between $6 to $42 billion US per year (Pendleton et al. 2012; Thomas 2014). Planning and investment decisions are based on direct financial benefits rather than broader environmental or economic concerns. A good example is shrimp farming where mangrove deforestation is a product of coastal aquaculture. Incomes from this industry range from $700 US per hectare to as much as $36,000 US per hectare with an average of about $6000 US. This is crucial because, in theory, ecosystem protection may be viable at moderate carbon prices to yield positive mitigation and adaptation outcomes; net economic returns on investment may be possible for as little as $15 to $20 US per hectare (Murray et al. 2011; Siikamäki et al. 2012). In practical terms, what this means is that to replace this income from farming, carbon payments would need to be at least $3.14 US per Mg CO_2 equivalents for low-profit farmers, $27 US for the average farmer and $156 US for high-income farmers (Yee 2010). The implication for blue carbon is that for a farmer to invest in conservation of resources rather than to continue farming, this option must have greater financial potential.

6.1.1 Payment for Ecosystem Services (PES)

Another one of the problems associated with blue carbon science and management is having a proper understanding of the actual cost of an ecosystem service, which is a tangible good or intangible function that benefits people. As pointed out by Lau (2013), coastal ecosystems simultaneously provide a number of services in addition to the potential for carbon storage. For instance, a mangrove forest or salt marsh may provide food, fuel, natural products, shoreline stabilisation, natural hazard protection, nutrient regulation (e.g. from storms, cyclones and floods), waste processing as well as supporting cultural services such as tourism, recreation, education spiritual values and aesthetics. In reality, these different services are all interconnected as well as interlinked to adjacent coastal ecosystems. A salt marsh, for instance, can provide some degree of protection from storm surges while also sequestering carbon and providing food (e.g. fish and shellfish) to locals as well as serving as a nursery ground for commercially valuable fisheries. Mangroves and seagrass meadows perform identical multiple functions that cannot be easily separated from one another. Further, their ability to perform such functions may depend upon the health of their own system, but also adjacent habitat, thus having cascading effects across ecosystems in the coastal zone.

With this service concept, ecosystem functions can be costed in terms of their ability to assist human well-being. For example, in 2003, coral reefs were estimated to provide $29.8 US billion annually in net benefits to humanity (Cesar et al. 2003). Tourism and recreation account for 32% of this value, coastal protection accounts for 30%, while fisheries and biodiversity account for 19%. Similarly, the World Resources Institute's Reefs-at-Risk programme estimates that the shoreline protection value of coral reefs and mangroves in Belize alone amounts to $231–$347 US million (Cooper et al. 2009), which approximates 9–14% of the nation's gross domestic product.

In Colombia, carbon sequestration benefits have been modelled into an economic system that has valued both mangrove and seagrasses within a new network of marine protected areas (Zarate-Barrera and Maldonado 2015). The model considers the capacity of mangroves and seagrasses for capturing and storing blue carbon and simulates scenarios for the variation of key variables, such as the market carbon price, the discount rate, the natural state of loss of these ecosystems and the expectations about the post-Kyoto agreements. The results of the model show that the expected benefits of blue carbon storage are substantial, but highly dependent on post-Kyoto negotiations and the dynamics of the carbon credit's demand and supply; natural loss rate of these ecosystems had no significant effect on the annual value of carbon stored. More importantly, under this scheme, the annual rates of carbon capture would increase from 49 to 94%, and total carbon storage would increase from 49 to 68% with respect to current protection areas.

A cost-benefit study has been done for mangrove plantations in northern Bohol in the Philippines to estimate the benefits, if any, of a 'win-win' scenario to mitigate climate change. Carandang et al. (2013) used three carbon prices in the international market to determine the net incremental benefits at different ages of mangrove plantations as well as net present values (**NPVs**) and prices of these plantations. They found that at the lowest price of $10 US per tonne the NPV is negative with it starting to become positive at a carbon price of $15US per tonne at year 20 up to year 50 with the corresponding computed NPV would be $167.16 US at year 20 and $467.14 US at age 50. All NPVs are positive once the carbon price reaches $20US per tonne. Therefore, establishing a carbon market for mangrove plantations is feasible, but very sensitive to the international carbon price. The additional problem would be the number of years of growth required to sustain mangrove carbon biomass during which time there is no guarantee what the carbon price on the international market will be.

While seagrass plantations do not yet exist, the sensitivity of payments for ecosystem services to carbon prices would also be an issue. Dewsbury et al. (2016) reviewed the prospects for further inclusion of seagrasses in climate policy frameworks as well as the potential for developing payment for ecosystem service (**PES**) schemes that are compatible with carbon management. They found that the prospects are slim, especially if targeted at the regulatory carbon market. This conclusion was reached mainly because of the doubts about the costs and financial markets and their relative instability. Voluntary carbon market schemes may be more promising, but these too are instable making a purely carbon market-based approach questionable, meaning that fluctuating carbon prices would impose excessive risk for a viable return on investment. Like mangroves and salt marshes, seagrass plantations or seagrass conserved areas would require a significant investment in time during which the international carbon price may fluctuate. What may seem as a solid investment at the start of a project may not be so solid several years later.

Some services (e.g. fisheries) are easier to estimate than others, and some are virtually impossible to estimate (e.g. cultural values). The problem is that there is no adequate 'one-size-fits-all' policy to determine the valuation of ecosystem services. As Lau (2013) points out, there are new policy tools and management mechanisms to correct for undervaluation and market failures, but at this stage, there is not even one overarching definition of payment for ecosystem services (**PES**).

Currently the valuation for PES is captured mostly for provisioning services such as fisheries while there is still a large gap in capturing the value of regulating, supporting and cultural services. Nevertheless, Lau (2013) has offered a framework for developing a PES scheme for blue carbon. First, clear identification of the ecosystem service in question (carbon sequestration) as well as the habitats where it is found and the biological and physical attributes contributing to provisioning of the ecosystem service is required. Second, the range of stakeholders who might be directly involved in the scheme should be identified. Third, the availability and suitability of performance indicators for baseline assessment and monitoring, the measurement of uncertainty and the management activities for achieving desired results need to be determined. For instance, the tonnes of CO_2 sequestered or in emissions avoided or carbon sequestration rates, the uncertainty of the methodologies used and proxy management activities such as prevention or reduction in deforestation/degradation are all issues that need to be considered in any PES scheme.

Few studies have estimated a monetary value for PES as it is difficult to so do for the reasons just discussed. However, Estrada et al. (2015) determined the value of mangrove carbon storage in south-eastern Brazil considering pre-existing estimates of carbon storage in the above-ground biomass and average transaction values of carbon credits. The mean monetary values ranged from \$19.00 US ha^{-1} year^{-1} for high intertidal basin forests to \$82.28 US ha^{-1} year^{-1} for low intertidal fringe forests. They estimated that the service of carbon sequestration may be worth up to \$455,827 US year^{-1} while carbon stored is worth \$3,477,041 US across all mangrove forests and values between \$104,311 and \$208,622 US ha^{-1} year^{-1} can be considered as the annual maintenance costs of this service.

The use of PES for coastal conservation via blue carbon appears feasible despite shortcomings. More research is required to elucidate the best practices to overcome these difficulties. For instance, more science and economics connecting specific management activities to produce a quantifiable outcome are necessary, as well as metrics and performance indicators to assess baselines and measure service delivery are required.

Also, new institutional frameworks will be required to manage payments and verify service delivery; education and capacity building will be required given the newness of such PES schemes. Payment for carbon credits is a clear outcome that can basically follow terrestrial PES schemes in including other ecosystem services (or at a minimum not excluding them). For example, managing a salt marsh to maximise carbon sequestration may not necessarily maximise the other ecosystem services. Perverse outcomes must thus be minimised. The key will be to identify those situations for which 'payments will be effective, cost-efficient, equitable and culturally acceptable, and those for which payments are not' (Lau 2013).

6.1.2 Regulatory and Policy Matters

Existing voluntary and regulated carbon markets are not equipped to address the complexities of social/ecological systems, for example, the problem of land tenure and traditional ownership. Markets do not recognise non-financial social and environmental benefits that might result in ecosystem-based carbon management. Projects are unlikely to proceed without providing goods and ecosystem services due to technical, institutional, administrative and financial constraints. Stakeholder engagement is required as blue carbon projects need to be commercially attractive propositions.

Insurance markets may be a way to finance or insure carbon products, that is, carbon stores are well-recognised as a market-based commodity. Certainly, governments and land owners have some incentive to insure their investment although there are constraints on the insurance pathway: lack of regulatory requirements, the absence of commercial incentives and resources, and physical practicalities (Thomas 2014). Thomas (2014) suggested that property insurance may be applied to blue carbon projects as carbon values and the wetlands themselves are already recognised in existing market-based instruments. A risk management approach involving a regulated or voluntary insurance instrument could be used to support a functional market for blue carbon.

The problem of time is a significant qualifier in any means to incorporate blue carbon projects into carbon markets. Plants take time to grow, and unlike commercial projects such as wheat, corn, barley, rice and rye, salt marsh grasses and seagrasses cannot be used as a commercial carbon product as nearly all of their carbon is stored in soils which take time to sequester significant and marketable amounts of carbon. Duarte et al. (2013a, b) recently examined the long-term potential of carbon sequestration in a seagrass restoration project by developing a model that combined models of patch growth, patch survival in seagrass planting projects and estimates of seagrass CO_2 sequestration per unit area for five seagrass species commonly used in restoration projects. They found that the cumulative carbon sequestered increased rapidly over time and planting density plateaued at 100 plants ha^{-1}. At this planting density, the modelled cumulative C sequestered ranges from 177 to over 1337 Mg CO_2 ha^{-1} over 50 years. The model thus suggests that the costs of seagrass restoration programmes may be fully recovered by the total CO_2 captured if there was a carbon tax in place in the given locale. Seagrass restoration programmes are therefore economically viable strategies to mitigate climate change through carbon sequestration.

The International Blue Carbon Policy Working Group has developed a blue carbon policy framework (Herr et al. 2012). Such a policy is timely as scientific understanding of wetland carbon capture and storage is sufficient to warrant development of effective policy, management and conservation incentives for coastal blue carbon. The development and implementation of blue carbon projects requires a policy framework that can deal with the management, conservation and financial issues arising from such a project. The policy framework was designed to:

- 'Define activities and a timeline to increase policy development, coastal planning and management activities that support and promote avoided degradation, conservation, restoration and sustainable use of coastal blue carbon systems;
- Define actions and a timeline to develop and implement financial and other incentives for climate change mitigation through conservation, restoration and sustainable use of coastal blue carbon;
- Identify key stakeholders, partners and blue carbon champions to implement the identified policy actions and define materials and products needed to support such activities; and

Table 6.1 Summary of the blue carbon policy framework

1. *Integrate blue carbon activities fully into the international policy and financing processes of the UNFCCC as part of mechanisms for climate change mitigation*
'Ensure recognition and inclusion of blue carbon sinks and sources into the outcome of the Durban Platform'
'Build awareness in the climate change policy community of the strength of scientific evidence of the carbon sequestered and stored in coastal ecosystems and of the emissions resulting from the degradation and destruction of these systems'
'Enhance the scientific and technical basis (data, reporting and accounting guidelines, methodologies, etc.) for financing of coastal carbon management activities'
'Access carbon finance through UNFCCC mechanisms and related funding streams'
'Include blue carbon management activities as incentives for climate change mitigation by Annex-I Parties'
'Monitor discussions on agriculture and its relevance for blue carbon'
'Support capacity-building activities to implement blue carbon management activities'
2. *Integrate blue carbon activities fully into other carbon finance mechanisms such as the voluntary carbon market as a mechanism for climate change mitigation*
3. *Develop a network of demonstration projects*
'Develop a strategic approach for the coordination and funding of demonstration projects'
'Provide capacity building at local and national level'
4. *Integrate blue carbon activities into other international, regional and national frameworks and policies, including coastal and marine frameworks and policies*
'Enhance implementation and inform financing processes of relevant Multilateral Environmental Agreements (**MEAs**) that provide policy frameworks relevant for coastal and marine ecosystem management'
'Use existing international frameworks to advance and disseminate technical knowledge on coastal ecosystems management for climate change mitigation'
'Use existing international frameworks to raise awareness of role of conservation, restoration and sustainable use of coastal ecosystems for climate change mitigation'
'Integrate coastal ecosystem conservation, sustainable use and restoration activities as a mechanism for climate change mitigation into relevant regional policy frameworks'
'Integrate coastal ecosystem conservation, sustainable use and restoration activities as a mechanism for climate change mitigation into existing national, subnational and sectoral policy framework'
5. *Facilitate the inclusion of the carbon value of coastal ecosystems in the accounting of ecosystem services*

From Herr et al. (2012)

- Identify opportunities, limits and risks of advancing blue carbon in different international climate, coastal and ocean policy fora'.

Table 6.1 summarises the five basic precepts for a policy framework on integrating blue carbon into the UNFCCC and other international and financing processes and markets. It also recommends a series of demonstration projects to begin the process of actually running a blue carbon project and bringing it to fruition. Blue carbon science and management need to be incorporated into international, regional and national frameworks that already exist to support climate change mitigation utilising coastal ecosystems, namely, wetlands such as salt marshes, mangrove forests and seagrass meadows.

The UNFCCC is the main mechanism by which blue carbon will be included into international frameworks. The UNFCCC in Article 4(d) calls for parties to 'promote sustainable management, and promote and cooperate in the conservation and enhancement, as appropriate, of sinks and reservoirs of all greenhouse gases not controlled by the Montreal Protocol, including [. . . . . .] oceans [. . . . . .] as well as [. . . . . .] other coastal and marine ecosystems'. As pointed out by Herr et al. (2012), coastal ecosystems have been largely excluded from UNFCCC-related mechanisms despite Article 4(d). However, a number of other mechanisms exist that currently support emission reductions and removals from natural systems under the UNFCCC: **REDD+** (Reducing Emissions from Deforestation and Forest Degradation), **NAMAs** (Nationally Appropriate Mitigation Actions) and **LULUCF** (Land-Use and some Land-Use Change and Forestry) including those implemented under **CDM** (Clean Development Mechanism). Blue carbon may therefore be included in these activities.

Outcomes from the new Durban Platform (a working group has been agreed under the 2011 UNFCCC COP17 meeting in Durban to address a variety of topics on mitigation, adaptation, finance, technology development and transfer, transparency of action and support and capacity building) have incorporated the contribution of natural carbon sinks and reservoirs to climate change mitigation and thus may include blue carbon activities.

Unfortunately, the climate change policy community is largely unaware of blue carbon research to date (Herr et al. 2012), so it is urgent that the level of awareness be highlighted to include the magnitude and strength of the ability of salt marshes, mangroves and seagrasses to sequester and store carbon and the danger of the continuing decline of the wetlands for GHG emissions.

Recently, the 2013 IPCC Guidelines for National GHG Inventories was completed with a chapter on coastal wetlands (Kennedy et al. 2014). This chapter has established data, reporting and accounting guidelines and levels of methods required to estimate national inventories of GHG emissions from salt marshes, mangrove forests and seagrass meadows. This chapter thus enhances the scientific and technical basis for financing of coastal carbon management activities.

To access carbon finance through UNFCCC mechanisms and related funding streams, Herr et al. (2012) recommend that (1) mangroves be incorporated into REDD+ activities as for terrestrial forests, (2) NAMAs be developed for coastal carbon ecosystems and (3) improved management of blue carbon coastal systems through climate change adaptation financing be supported. Capacity building also needs to be supported as it is essential for developing nations to have the ability to conduct and manage their own blue carbon projects. It has also been pointed out that other carbon finance mechanisms can be used to fund blue carbon projects, such as current organisations like the Verified Carbon Standard (**VCS**) or the American Climate Registry (**ACR**) which are used by carbon mitigation projects to verify and issue carbon credits for the international voluntary offset market.

A network of demonstration projects is needed to show the viability of blue carbon and to work out in a practical way the problems and pitfalls of running a project. Demonstration projects will provide a venue for testing methodologies and for testing tools for the UNFCCC and other frameworks that support carbon accounting. Capacity building is also a good reason for demonstration projects as they are essentially a teaching tool for national abilities to work and run a project. The most challenging problem for demonstration projects is getting the initial funding.

Blue carbon needs to be integrated into international, regional and national frameworks and policies, and there are a number of policy frameworks that already make reference to conservation, sustainable use and restoration of, and reduced emissions from, coastal ecosystems: the Convention on Biological Diversity (**CBD**), Ramsar Convention on Wetlands (**RAMSAR**), UN Conference on Sustainable Development (Rio +20), UN Open-ended Informal Consultative Process on Oceans and the Law of the Sea and UNEP Global Programme of Action for the Protection of the Marine Environment from Land-based Activities (**GPA-Marine**). Meetings and communications associated with policy frameworks will provide an opportunity for building awareness and support for coastal blue carbon.

Blue carbon as a vehicle for conservation, sustainable use and restoration also needs to be fully integrated into existing national and regional policy frameworks as a mechanism for climate change mitigation. This may be time-consuming and difficult as only some developing nations have a national or a series of subregional policies on blue carbon or coastal ecosystem use in climate change mitigation. Perhaps the best way to accomplish integration is to communicate the strength of the coastal carbon sinks and how wetlands link closely into existing frameworks on policies for watersheds, including agriculture and flood control. Another pathway to integration is the insurance industry, which already recognises the value of coastal habitats in protection against storm damage, sea-level rise and flooding risk. Integration may also be forthcoming in the aquaculture industry when the emissions from aquaculture are offset by the savings from conserving remaining habitat; we have seen in earlier chapters how great GHG emissions are as a result of habitat destruction. This pathway may also take advantage of stacking and bundling of ecosystem services as, for example, remaining mangrove forests in close proximity to a shrimp farm may still retain a nursery function and other functions such as shoreline protection, water clarity and a source of biodiversity.

These later functions are well-established for blue carbon wetlands, but vulnerability assessments are still needed involving basic science parameters of the ecosystem as well as local community knowledge. There is also the need to highlight the critical role of social factors in vulnerability assessment and development planning as well as the need to incorporate other ecosystem services such as maintaining biodiversity and storm protection. However, one of the main problems with vulnerability assessments traditionally is that they tend to focus solely on sea-level rise without considering other aspects of climate change. Osland et al. (2016) make the argument that macroclimatic drivers (temperature, rainfall) need to be considered in vulnerability assessments as they are for terrestrial ecosystems. They show how even small changes in macroclimatic conditions can foster large changes in wetland ecosystem structure and function.

6.2   Restoration and Management

A number of blue carbon projects have been and are currently operating, mostly recently as demonstration projects involving the rehabilitation and restoration of mangrove forests (Table 6.2). Most are capacity-building exercises and financed from research or public development institutions; few are private sector projects through generation of carbon offsets for the voluntary/regulated carbon market. No project as yet has sold carbon offsets to market illustrating that at this point in time blue carbon is either not yet a viable market commodity or communication with private investors is lacking. It is also true to say that blue carbon is still a new initiative and it will take some time to generate private investment until these demonstrations projects provide 'proof of concept'. Wylie et al. (2016) describe the tools necessary to make a successful blue carbon project by examining case studies; there are benefits in (1) incorporating livelihood aspects as part of the project and (2) involving members of the local community in all stages of planning and implementation. The importance of involving local communities is common sense as a project cannot succeed in isolation. Community involvement ensures that 'leakage' does not occur, that is, protection in one place does not lead to destruction someplace else. Blue carbon projects may not be able to overcome the threats that will likely occur due to local use of, for example, mangroves unless the local community sees benefit such as opportunities for income. Wylie et al. (2016) argue that there is much benefit in small, community-based projects in financing via the voluntary carbon market as the requirements are less stringent than financing via UNFCCC mechanisms.

The reader is referred to the following websites for updates on current and future projects: thebluecarbonproject.com,   thebluecarboninitiative.org   and blucarbonportal.org.

6.2.1   Success or Failure: What Does and Doesn't Work

The current projects are small-scale and focused on science or economics with little if any merging of the two, and it will take some time to fully integrate them into the conservation and management sphere, at least not until they achieve 'proof of concept' successfully. In fact, a perusal of the literature indicates that most restoration and rehabilitation projects, especially of salt marshes and seagrass beds, have failed to meet success criteria. Landscape setting, habitat type, hydrological regime,

Table 6.2 Some blue carbon projects past and present around the globe

Location	System	Activity	Proponents	Project type	Financial return	Other outputs
Brazil	S	DC	Instituto de Oceanografia-Federal University of Rio Grande	R	None	Biophysical data including carbon dynamics
Brazil	S	DC	Universidade Estadual de Rio de Janeiro; Universidade Federal of Rio Grande; Universidade Federal de Santa Catariana e Universidade Federal Rural de Pernambuco	R	None	Biophysical data including inferred carbon stocks
Brazil	SM, S, M	DC	Institute of Oceanography-University of Sao Paulo with collaboration of 39 Brazilian institutions	R	None	Biophysical data including inferred carbon stocks
China	M	DC	Tsinghua University; Xiamen University	R	None	C sequestration and flux baseline data; social and demographic data
Tanzania	M	DC	WWF; Sokoine University of Agriculture; University of Dar es Salaam; Lawyers Environmental Action Team; Journalists Environment Team	R	None	C sequestration and flux baseline data; REDD+ policy integration
USA	SM	R	University of Maryland; US Fish and Wildlife Service	R	None	C sequestration and flux baseline data
	M	R, AE		PES	None	

(continued)

Table 6.2 (continued)

Location	System	Activity	Proponents	Project type	Financial return	Other outputs
Democratic Republic of the Congo, Cameroon			UNEP; Cameroon Wildlife Conservation Society; UNEP-World Conservation Monitoring Centre; Kenya Marine and Fisheries Research Institute			C sequestration and flux baseline data; REDD+ policy integration
Madagascar	M	R, AE	Blue Ventures	PES	None	C sequestration and flux baseline data; REDD+ policy integration
Gambia, Guinea and Guinea-Bissau	M	R, AE	UNEP, Canary Current Large Marine Ecosystem; Wetlands International; IUCN	PES	None	Biophysical data including carbon dynamics
Indonesia	M	R	Wetlands International; The Nature Conservancy; Deltares; Wageningen University; various Indonesian partner organisations	PES	None	Research and publication of 'Mangrove Capital' to guide planning and development
Costa Rica	M	DC	Tropical Agricultural Research and Higher Education Center; BIOMARC Project; Universidad Nacional de Costa Rica	PES	None	C sequestration and flux baseline data; community participation and capacity building
Abu Dhabi, UAE	M	DC	Abu Dhabi Global Environmental Data Initiative (AGEDI)	BC	None	C sequestration and flux baseline data; REDD+ policy integration

(continued)

Table 6.2 (continued)

Location	System	Activity	Proponents	Project type	Financial return	Other outputs
Philippines	M, S	DC	Science & Technology Research Partnership for Sustainable Development; JICA-JST; Tokyo Institute of Technology	BC	None	C sequestration and flux baseline data; REDD+ policy integration
USA	SM	AE	Waquoit Bay National Estuarine Research Reserve; NOAA; National Estuarine Reserve Research System Science Collaborative	BC	None	C sequestration and flux baseline data; carbon stock assessment tool
Vietnam	M	AE	SNV Netherlands, IUCN, International Climate Initiative, German Federal Ministry for the Environment, Building and Nuclear Safety (BMU), Minh Phu Liveihoods, Danone Fund for Nature	BC	Premium market price from shrimp while conserving mangroves	Financing from Naturland Organic Shrimp Certification
Panama	M	AE	UNDP; Panama Environment Authority; Panama Aquatic Resources Authority; The Nature Conservancy; Wetlands International	BC	None	C sequestration and flux baseline data; REDD+ policy integration

(continued)

Table 6.2 (continued)

Location	System	Activity	Proponents	Project type	Financial return	Other outputs
Kenya	M	R, AE	Napier University; Kenya Marine Fisheries Institute; Earthwatch Institute	BC	Expected to generate 2.5 kt CO_2e/year or $12,000 for 20 years	Registered small-scale Plan Vivo (voluntary scheme) restoration project (see technical details at Plan Vivo Foundation website
Senegal	M	R	Livelihoods Fund; L'Oceanium de Dakar	BC	Expected to generate 2.7 kt CO_2e/year for 30 years	Registered CDM small-scale reforestation project (see cdm.unfccc.int project ref#5265)
Mozambique	M	AE	WWF; US Forest Service; USAID; University of Eduardo Mondlane; Kenya Marine and Fisheries Research Institute	BC	None	Biophysical data including carbon dynamics
Ghana	M	R	Coastal Resources Center	BC	None	C sequestration and flux baseline data; REDD+ policy integration
India	M	R	Livelihoods Fund	BC	Expected to generate 8 kt CO_2e/year for 20 years	CDM small-scale afforestation/reforestation project (see cdm.unfccc.int) in the Sundarbans
China	S, M, SM	DC	Tsinghua University; Xiamen University; State Oceanic Administration	BC	None	C sequestration and flux baseline data
Indonesia	M	R	Ministry of Forestry of Batam City; Y.L. Invest Co; Team Permanent Mangrove	BC	Expected to generate 3.8 kt CO_2e/year for 30 years	CDM small-scale afforestation/reforestation project (see cdm.unfccc.int)

(continued)

Table 6.2 (continued)

Location	System	Activity	Proponents	Project type	Financial return	Other outputs
Indonesia	M	R	Livelihoods Fund; Yagasu Aceh	BC	Expected to generate 105 kt CO_2e/year for 20 years	VCS reforestation project
Indonesia	S, M	R, AE	Agency for Research & Development of Marine & Fisheries; Ministry of Marine Affairs; Fisheries Indonesia	BC	None	C sequestration and flux baseline data; REDD+ policy integration
Indonesia	M .	R, AE	Charles Darwin University; Japesda; Yayasan Hutan Biru	BC	None	Ecological restoration; community development
Indonesia	M	R	Wetlands International	BC	None to date	Community-based microcredit programme to improve shrimp farming through mangrove restoration with carbon credits produced

Updated from Thomas (2014)

Abbreviations: *SM* salt marsh, *S* seagrass, *M* mangrove, *RST* restoration, *DC* data collection, *AE* avoided emissions, *R* research, *PES* payment for ecosystem services, *BC* blue carbon

soil properties, invasive species, disturbance regimes, seed banks and declining biodiversity among a host of factors can constrain the restoration process (Zedler 2000).

There is a problem in that most restoration and rehabilitation projects have suffered from poor management protocols such as not having proper success or failure criteria and have suffered from poor or uncertain methodology. There are few if any clear guidelines for restoration of seagrasses (van Katwijk et al. 2009), mangroves (Field 1998; Ellison 2000; Wylie et al. 2016) and salt marshes (Williams and Faber 2001). The experiences of salt marsh restoration in San Francisco Bay (Williams and Faber 2001) indicate several important issues learned that are also applicable for mangrove and seagrass restoration:

- 'Habitats can be restored if the correct sites have been chosen;
- Methodology of restoration is still experimental as it is not known what percentage of the original ecosystem function returns nor how long it takes;
- Successful restoration is greatly dependent on restoration of hydrodynamic processes;
- Restoration projects must have clear statements of measurable, achievable biological objectives including success and failure attributes;
- Restored habitats are best viewed as immature ecosystems that will mature with time;
- Natural evolution of ecological processes of a restored habitat may take a long period of time;
- Monitoring of restoration is mandatory in order to determine the success or failure of the project including the amount of carbon sequestered;
- Planning and management of physical processes should preferably be on the conservative side to allow for habitat development'.

For mangrove ecosystems, Lewis (2005) reviewed the existing information as well as his own practical work and concluded that assessing the existing hydrology of natural habitats and then applying this information to a habitat to be restored is of prime importance. His restoration principles:

1. 'Get the hydrology right first;
2. Find out why a given site has lost its mangroves or why the given site has never had mangroves;
3. Once you find out why, see if you can correct the conditions that currently prevents natural colonization of the selected mangrove restoration site. If you cannot correct these conditions, pick another site;
4. Use a reference mangrove site for examining normal hydrology for mangroves in your particular area. . . . . . establish the same range of elevations as your reference site at the site to be restored or restore the same hydrology to an impounded mangrove by breaching the dikes in the right places. The "right places" are usually the mouths of historic tidal creeks. These are often visible in. . . . . photographs;
5. Remember that mangrove do not have flat floors. There are subtle topographic changes that control tidal flooding depth, duration and frequency. Understand the normal topography of your reference forest before attempting to restore another area;
6. Construction of tidal creeks within restored mangrove forests facilitates flooding and drainage, and allows for entrée and exit of fish (and other biota and nutrients) with the tide: and
7. Evaluate costs of restoration early in project design to make your project as cost effective as possible'.

These principals are also valid for salt marshes and, with some adaptation, for seagrass beds. The first European Seagrass Restoration Workshop concluded similarly that priority should be given to natural restoration, with emphasis on the fact that 'restoration should never be considered the first alterative when planning for the mitigation of coastal development projects or to justify mitigation as a compensation measure for economic activities' (Cunha et al. 2012). The results show that none of the seagrass restoration projects developed in Europe by the participants during the past 10 years was successful. The group endorsed several recommendations prior to the start of a restoration project:

1. 'Establish clear goals and objectives prior to initiation of restoration;
2. Define monitoring methods and success criteria...... and make accommodations for long-term monitoring (i.e. 5–10 years) a part of the initial project;
3. Include donor population monitoring in the project;
4. Make every effort to ensure that local threats (e.g., bioturbation, herbivory, hydrology, sediment movements, human impact, etc.) to seagrasses are well known........start only when all threats causing loss have been eliminated;
5. Initiate with small-scale or pilot restoration trials...........;
6. Devices to anchor plants or protect them against storms, sediment dynamics or herbivory should be avoided......;
7. Covering the transplant rhizomes with a local stone or sand bag to improve the technique seems to be a positive exception...............provided that sites are carefully selected.......;
8. The application of a shell layer is another positive exception as it works to stabilize sediments......;
9. Traditional local knowledge can give a big help..............spread the trials throughout different sites and use different methods. Learn and be willing to change plans based on the experienced results (adaptive management);
10. Strive to learn from the experience of others and use the information to improve methods at different sites. It seems that it may take more than 5–10 years to start becoming successful;
11. Almost all scientists expressed frustration about natural beds being disturbed and/or natural recovery being prevented (trawling, shellfish/bait collection, tourist activities, etc.). This is partly due to the absence of law enforcement and partly due to limited regulation or protection status or modification of protection status if economics prevail. Make sure you have identified all these constraints and their magnitude and frequency before starting a restoration effort'.

Success criteria for seagrass restoration in many areas focuses on persistence, area restored and shoot density (Fonseca et al. 2000). Van Katwijk et al. (2009) concluded that the success of a seagrass restoration project is dependent on habitat selection and selection of the donor population, spreading of risks and ecosystem engineering efforts. This is also true for salt marshes and mangroves. For example, Arachchilage et al. (2017) found in assessing restoration efforts of mangroves in Sri Lanka that restoration success is highly variable, with success rates varying from 0 to 78%; 9 of 23 project sites showed no surviving plants.

How do we assess the success of a restoration project in terms of blue carbon? Carbon storage as a result of the project can be calculated by measuring the addition of blue carbon in the restored site by measuring the C_{org} content in soils multiplied by some measure of recent sedimentation, for example, by R-SET. Marbá et al. (2015) reconstructed the trajectories of carbon stocks associated with one of the longest monitored seagrass restoration projects. They demonstrated that sediment carbon stocks erode following seagrass loss and that revegetated projects restore seagrass carbon sequestration capacity by combining carbon chronosequences with ^{210}Pb dating of seagrass sediments in a meadow that experienced losses until the end of the 1980s and subsequent serial revegetation efforts. Inventories of excess ^{210}Pb showed that its accumulation and thus sediments coincided with the presence of seagrass vegetation. Seagrass regeneration enhanced carbon deposition and burial with carbon burial rates increasing with the age of restored sites; 18 years after planting, they were similar to that in continuously vegetated beds. Greiner et al. (2013) similarly found that seagrass restoration enhances carbon sequestration. In their study of meadows of different age in Virginia, measurements were made of percent carbon and ^{210}Pb from dating at 1 cm intervals to a depth of 10 cm. They found that carbon accumulation rates were higher in 10-year-old meadows compared with 4-year-old beds and bare sediment.

Can coastal ecosystems be managed to sequester more carbon? Macreadie et al. (2017a; b) discussed three potential management strategies that hold some promise for optimising carbon sequestration:

1. 'Reducing anthropogenic nutrient inputs;
2. Reinstating top-down control of bioturbator populations; and
3. Restoring hydrology'

The first management strategy is true in that most evidence shows that there is a decrease in carbon storage with nutrient addition. For example, there are usually net losses of carbon either through plant mortality and gaseous efflux or through erosion and loss of sediment. A risk assessment by Lovelock et al. (2017) has shown that there is increased risk of high CO_2 emissions in blue carbon ecosystems with increasing stocks of soil organic carbon. The second strategy is based on evidence that shows that high densities of bioturbators can have negative impacts on soil carbon stocks and fluxes; low to moderate levels of bioturbation help stimulate plant growth, but high levels result in high losses of CO_2. The third strategy involves the reestablishment of tidal exchange which will modulate CO_2 fluxes back to natural rates of emission. Data from ponded systems has shown that conversion of coastal ecosystems through tidal flow restriction can disrupt carbon sequestration by coastal ecosystems and may switch these ecosystems from being net sinks to net sources of carbon (Lovelock et al. 2017).

6.3 Financing

Financing remains a key concern for blue carbon projects to eventuate and proceed successfully. Table 6.3 summarises the range of different funding approaches to blue carbon projects in developing and developed nations with the type of finance and whether or not the carbon benefit flows or remains.

Table 6.3 Features of different funding approaches to blue carbon activities

Activity	Can occur in a developing (D) or developed (DV) country	Finance	Carbon benefit
NAMAs/NAPAs[a]	D	DO, I	R
Climate-related ODA[b]	D	I	R
Bi- and multi-lateral activities[c]	D, DV	DO, I	R
REDD+	D	I	F
National NRM actions[d]	D, DV	DO	R
Voluntary offsets (e.g. VCS)[e]	D, DV	P	R, F
Compliance offsets (e.g. CDM)[f]	D	P	F
Domestic compliance offsets (e.g. CFI, CCERs)[g]	D, DV	P	R
CSR projects[h]	D, DV	P	R
Others (insurance microfinance, green bonds)[i]	D, DV	DO, I, P	R

Modified from Thomas (2014)

Abbreviations: *DO* domestic public finance, *I* International public finance, *P* private, *R* remains, *F* flows

[a]Nationally Appropriate Mitigation Actions (NAMAs) are agreed actions taken by developing countries as part of their commitments under the terms of the UNFCCC. National Adaptation Programmes of Action (NAPAs) are limited to least developed countries

[b]Official development assistance (see www.oecd.org/dac/)

[c]Bi- and multilateral activities refer to agreements made between nations or regional groups of nations or activities implemented through partnerships with public funding institutions such as the World Bank or Asian Development Bank

[d]Natural resource management (NRM) at the national level can occur in a variety of ways depending on local regulatory and social conditions

[e]Voluntary market carbon offsets can be sourced through a variety of providers including Verified Carbon Standard, the American Carbon Registry and others. China has created its own domestic carbon offset

[f]Regulated domestic emissions trading schemes require international carbon offsets to be sourced from benchmark mechanisms, principally the Clean Development Mechanism (CDM) and joint implementation (JI) schemes established by the Kyoto Protocol to the UNFCCC

[g]National carbon reduction compliance schemes continue to be established, and these legislative initiatives usually create their own unique domestic carbon offset units, generally oriented towards eventual integration with international market mechanisms

[h]Corporate social responsibility (CSR) is an important area for potential blue carbon funding that may not be considered in most discussions of climate finance opportunities, because many large organisations may choose to invest in voluntary projects without a carbon focus. Depending on the scale of the activity, this might be a useful consideration for project developers

[i]Climate bonds are a new class of financial asset that can be issued by governments or private institutions and operate in the same manner as standard debt instruments. Climate bonds may be a model for new classes of asset including insurance projects. Essentially, funding can come from three types of sources: (1) national government, (2) development of pilot programmes and (3) payment from verified emissions reductions, that is, carbon offset schemes, as under the UNFCCC nations agree to individual emission reduction commitments which can be achieved through three flexible mechanisms: (1) international emissions trading, (2) joint implementation and (3) Clean Development Mechanism

Herr et al. (2015) have recently reviewed finance mechanisms for blue carbon projects. They point to an increasing interest by governments, NGOs, local communities and academia to support coastal wetland restoration and conservation, but observe that finding appropriate funding to set up such a blue carbon project or to develop a national scheme for blue carbon remains 'a challenge'.

6.3.1   UNFCCC-Related and Other Finance Mechanisms

As noted earlier, the UNFCCC sets the framework for internationally agreed GHG reduction measures and provides technical details and funds to support a variety of climate mitigation activities including carbon mechanisms. Specific financial mechanisms under the UNFCCC umbrella include the **GEF** Trust Fund, the Special Climate Change Fund (**SCCF**), the Least Developed Countries Fund (**LDCF**), the Green Climate Fund (**GCF**) and the Adaptation Fund. Other multilateral and national climate funds include the BioCarbon Fund and other funds from the African and Asian Development Banks.

The list of possible sources is confusing and has been described as a 'jungle' (Herr et al. 2015). Herr et al. (2015) show how to start looking for funds within the multiple funding agencies. First, one needs to determine the type of activity, that is, whether it is starting up a national programme, subnational programme or an individual blue carbon project. Second, one needs to match up with a possible funding source, for example, in this case the Green Climate Fund or REDD+; national funds are also available such as IKI, NEFCO and GCPF. Third, development banks do provide funds for mid- (<$2 million US) to full-size (>$2 million US) projects although projects in this size range usually require government support. Fourth, one needs to decide on whether or not incremental or additional funds are necessary. For example, if biodiversity is a supplemental issue there are RAMSAR Small Grants or other sources such as biodiversity funds from development banks. Small projects under $500,000 US can fit well with foundations, charities or the private sector. Mid- and full-size projects are best funded by UNFCCC-related sectors such as the Global Environment Fund (**GEF**).

Some financing is best suited to specific habitat. For example, although the REDD+ financing mechanism is still being arranged, mangroves are well suited for REDD+ financing, being forests with similar ecological traits to terrestrial forests (Yee 2010; Ahmed and Glaser 2016; Mashayekhi et al. 2016). The main funding streams are those of the Forest Carbon Partnership Facility (FCPF) of the World Bank and UN-REDD. The former is a global partnership of governments, business, society and indigenous peoples and is broken up into two separate but complimentary funding mechanisms: The Readiness Fund and the Carbon Fund. Currently, there are 47 participating countries in these programmes (Herr et al. 2015). The UN-REDD programme is a collaboration among the UNDP, FAO and UNEP and supports national initiatives in 64 partner countries (Herr et al. 2015). Herr et al. (2015) list relevant online sites for available climate adaptation and mitigation funding.

6.3.2 The Voluntary Carbon Market

Blue carbon projects can also be funded via the voluntary carbon market. A good example of this type of project is in Madagascar (Table 6.2) and has been run by Blue Ventures since 2011. The project has two demonstration sites, one a large-scale (26,000 ha) mangrove project and the second a smaller project (1015 ha). Both are being used to test the feasibility of using blue carbon as a long-term financial mechanism for community-based mangrove management.

One of the pitfalls of the voluntary carbon market is that the price of carbon fluctuates over time, and this may affect the viability of a blue carbon project. For instance, Jerath et al. (2012) noted that the social cost of carbon (SCC) ranges from $9 US to $50 US per tonne of carbon while marginal abatement costs (MACs) vary from $70 US to $616 US per tonne of carbon. Both SCC and MACs are useful for setting a price for carbon in the absence of efficient carbon markets. Carbon prices also vary across countries and markets, and people's willingness to pay is expected to correspondingly increase with their view that carbon storage will provide significant profit.

The voluntary carbon market deals with the selling and buying of emission reduction credits (offsets) in non-government-regulated markets. The demand for verified carbon credits is market-driven, that is, by customer demand. There are many types of buyers in the market, from individuals who want to offset their carbon footprint from air travel to companies who themselves emit GHGs. Companies do this to enable themselves to be labelled clean and green. As discussed earlier, coastal carbon offset projects may be economically feasible at low to moderate carbon prices of $2 to $11 US per tonne CO_2-e. The majority of potential emissions from mangroves could be avoided at less than $10 US per tonne CO_2-e (Siikamäki et al. 2012).

Efforts are currently underway to develop methodologies for verifying coastal carbon credits. The Verified Carbon Standard (VCS) and American Climate Registry (ACR) are used globally to verify and issue carbon credits from field projects such as the one in Madagascar to be traded on the voluntary carbon market. Other standards include The Climate, Community and Biodiversity Standard (CCB), the CarbonFix Standard and the Plan Vivo Systems and Standard. Obviously, a blue carbon project that is going into the voluntary carbon market needs to find an appropriate standard as well as methodologies to measure, report and verify changes in carbon storage although no verified standard organisations have yet produced such accepted procedures.

Biodiversity can also be a focus of funding opportunities, as noted above. The Ramsar Convention maintains three direct assistance programmes: the Small Grants

Fund, the Wetland for the Future capacity-building programme and the Swiss Grant for Africa. These funds may be tapped into for a blue carbon project, but an analysis of cost-effectiveness still needs to be done for projects. Adame et al. (2015a; b) suggested using Marxan, a spatial prioritisation tool to balance the provision of ecosystem services versus the cost of restoration. Their approach efficiently selected restoration sites that at low cost were compatible with biodiversity targets; the restoration of biodiversity was largely guaranteed by choosing areas for restoration based on the potential for carbon storage.

Debt-for-nature swaps can also be an innovative, non-market way of financing. A debt-for-nature swap involves a lending country selling the debt owed by a recipient country (the debtor) to a third party at less than the full value of the original loan. In exchange the indebted country agrees to a payment schedule on the amount of the debt remaining. The third party then uses the debt repayments to support domestic conservation initiatives. An example of this type of funding mechanism is in the Seychelles where there has been a debt swap for conservation and adaptation (Herr et al. 2015). In this project, the Seychelles Debt Swap for Conservation and Adaptation between the Seychelles government and the Club of Paris developed through the platform of the Global Island Partnership with the technical support of The Nature Conservancy (TNC) develops a long-term funding stream for conservation activities.

Another pathway, as noted earlier, is via payment for ecosystem services. An example of this type of arrangement is in Ecuador where mangroves are held under preservation and protection agreements; by late 2018, it hopes to have 100,000 ha of mangrove forest under protection via a mix of fixed and variable payments. The fixed yearly payment amounts to $7000 US for areas between 100 and 500 ha, $10,000 US for areas between 501 and 1000 ha and $15,000 US for areas above 1000 ha. Variable payments depend on the size of the area as well amounting to a benefit of $3 US ha^{-1} year^{-1}. The PES schemes nonetheless offer the greatest scientific and policy challenges as accurate valuations will offer incentives for funding and private investment as well as improve management and governance of these resources.

There are problems with valuations that remain difficult to solve except on a case-by-case basis. First, there are a large number of services that are interlinked thus making it difficult to value one particular service. Second, as Bardesgaard (2016a; b) has noted, the commoditisation of nature might encourage perverse outcomes and represents a shift from conservation motives to economic self-interest, that is, the expectation of financial returns from investment. Third, valuation will depend on the rate of habitat loss; if current trends continue, less carbon will be sequestered, leading to a decline in value (Beaumont et al. 2014). An alternative is to consider the quality of environmental assets rather than ecosystem services. Quality assessments can then be quantitatively assessed (e.g. species richness, habitat quality, cultural values).

6.3.3 Investment Risk

Investment is all about risk. Risks need to be minimised in order to maximise the probability of a return on the investment. There is also a need to demonstrate the likelihood of attractive returns (Warner et al. 2016). This idea is constrained by (1) biophysical issues such as amount of carbon sequestered, measurement uncertainty and logistical challenges; (2) technical capacity and infrastructure; (3) concerns over governance (corruption, land tenure); (4) existence of regulatory frameworks; (5) permanence of the ecological asset; (6) security of property tenure; (7) the temporal scale of measurements (i.e. ecosystems need time to mature for increased carbon storage); and (8) the fact that there may need to be different policy instruments designed for the specific type of finance (e.g. publicly funded versus private investment).

6.3.4 Policy and Commodification

The science policy and management community have a few naysayers regarding blue carbon, and these problems must be addressed fully before blue carbon can mature as a viable business proposition. Broadhead (2011) cited the difficulty in marketing many ecosystem services that mangroves can provide as well as the lack of clarity over ownership of natural ecosystems. In almost all cases, the value of goods and services produced by mangroves has not been fully realised. In addition, the difficulty of realising the non-market benefits of mangroves is compounded by the fact that the benefits accrue to many people most of whom are poor. Broadhead (2011) further maintains that conversion of mangroves is generally associated with a change in ownership towards an individual or 'an established entity' while benefits are not commonly accessed across the local community. A range of numerous other problems need to be sorted, such as technical considerations associated with monitoring and quantifying carbon flow with precision. Also, setting baselines have meant that costs associated with these issues may exceed benefits.

Concerns raised by Barbesgaard (2016a, b) focus on the concept of 'ocean grabbing' in which private industry takes through ownership what is essentially common property. It is pointed out that social movements have called blue carbon projects a 'false solution' because of what may be the false belief that market logic provides the best tool to organise society and conserve nature. Commodification of nature involves large shifts in and struggles over social relations such as ownership of natural resources, socio-economic inequality and power. Under the blue carbon umbrella nature is reduced to a commodity to buy and sell violating the ideals of social justice. Barbegaard (2016a, b) points out that 'blue carbon projects act as a smoke-screen diverting attention away from the systematic changes needed to stop the climate crisis [. . . ..] polluting actors, be they states or transnational corporations [. . . ..] can continue to pollute and destroy one place as long as a coastal ecosystem that stores and absorbs carbon somewhere else is 'protected'. This idea is not a new argument as it originates from the old idea of corporations' land grabbing. The 21st Conference of the Parties to the UNFCCC closed with the initiation on the Paris Agreement on 12 December 2015 with a high level of collaboration from nations and corporations. However, radical transnational agrarian and social justice movements argue that the agreement will facilitate continued market-based resource grabs for land, forests and oceans through carbon trading schemes and related mechanisms (Tramel 2016). While there may be an element of truth in that perverse outcomes can eventuate, a proper policy framework that included safeguards can overcome these concerns.

References

Adame MF, Hermoso V, Perhans K, Lovelock CE, Herrerea-Silveira JA (2015a) Selecting cost-effective areas for restoration of ecosystem services. Conserv Biol 29:493–502

Adame MF, Santini NS, Tovilla C, Vázquez-Lule A, Castro L (2015b) Carbon stocks and soil sequestration rates in riverine mangroves and freshwater wetlands. Biogeosci Discuss 12:1015–1045

Ahmed N, Glaser M (2016) Coastal aquaculture, mangrove deforestation and blue carbon emissions: is REDD+ a solution? Mar Pol 66:58–66

Arachchilage K, Kodikara S, Mukherjee N, Jayatissa LP, Dahdouh-Guebas F, Koedam N (2017) Have mangrove restoration projects worked? An in-depth study in Sri Lanka. Restor Ecol 25:705–716

Barbesgaard MC (2016a) Blue growth: saviour or ocean grabbing? Global governance/politics, climate justice & agrarian/ social justice: linkages and challenges, an international colloquium 4–5 February 2016. Colloquium paper No. 5. International Institute of Social Studies, The Hague

Barbesgaard MC (2016b) Blue carbon: ocean grabbing in disguise? Issue brief. Transnational Institute, Afrika Kontakt, Indonesia Traditional Fisherfolks Union, Amsterdam/Copenhagen/ Jakarta

Beaumont NJ, Jones L, Garbutt A, Hansom JD, Toberman M (2014) The value of carbon sequestration and storage in coastal habitats. Estuar Coast Shelf Sci 137:32–40

Broadhead JS (2011) Reality check on the potential to generate income from mangroves through a carbon credit sales and payments for environmental services. Regional Fisheries Livelihoods Programme for South and Southeast Asia (RFLP). FAO, Rome

Carandang AP, Leni D, Camacho D, Gervana T, Dizon JT, Camacho SC, deLuna CC, Pulkin FB, Combalicer EA, Paras FD, Peras RJJ, Rebugio LL (2013) Economic valuation for sustainable mangrove ecosystem management in Bohol and Palawan, Philippines. Forest Sci Technol 9:118–125

Cesar HJS, Burke L, Pet-Soede L (2003) The economics of worldwide coral reef degradation. Cesar Environmental Economics Consulting/WWF-Netherlands, Arnhem /Zeist

Cooper E, Burke L, Bood N (2009) Coastal capital: Belize. The economic contribution of Belize's coral reefs and mangroves. World Resources Institute, Washington, DC

Cunha AH, Marbá NN, van Katwijk MM, Pickerell C, Henriques M, Bernard G, Ferreira MA, Garcia S, Garmendia JM, Manent P (2012) Changing paradigms in seagrass restoration. Restor Ecol 20:427–430

Dewsbury BM, Bhat M, Fourqurean JW (2016) A review of seagrass economic valuations: gaps and progress in valuation approaches. Ecosystem Serv 18:68–77

Duarte CM, Losada IJ, Hendriks IE, Mazarrasa I, Marbá N (2013a) The role of coastal plant communities for climate change mitigation and adaptation. Nat Climate Change 3:961–968

Duarte CM, Sintes T, Marbá N (2013b) Assessing the CO_2 capture potential of seagrass restoration projects. J Appl Ecol 50:1341–1349

Ellison AM (2000) Mangrove restoration: do we know enough? Restor Ecol 8:219–229

Estrada GCD, Soares MLG, Fernandez V, de Almeida RMM (2015) The economic valuation of carbon storage and sequestration as ecosystem services of mangroves: a case study from southeastern Brazil. Ecosyst Serv Manage 11:29–35

Field CD (1998) Rehabilitation of mangrove ecosystems. Mar Pollut Bull 37:383–392

Fonseca MS, Julius BE, Kenworthy WJ (2000) Integrating biology and economics in seagrass restoration: how much is enough and why? Ecol Eng 15:227–237

Greiner JT, McGlathery KJ, Gunnell J, McKee BA (2013) Seagrass restoration enhances "blue carbon" sequestration in coastal waters. PLoS ONE 8:e72469

Herr D, Pidgeon E, Laffoley D (eds) (2012) Blue carbon policy framework: based on the discussion of the International Blue Carbon Policy Working Group. IUCN, Gland/Arlington

Herr D, Agardy T, Benzaken D, Hicks F, Howard J, Landis E, Soles A, Vegh T (2015) Coastal "blue" carbon. A revised guide to supporting coastal wetland programs and projects using climate finance and other financial mechanisms. IUCN, Gland

Jerath M, Mahadev GB, Rivera-Monroy VH (2012) Alternative approaches to valuing carbon sequestration in mangroves. Proc ISEE 2012 Conf Ecol Econ Rio 20:156–165

Kennedy H, Alongi DM, Karim A, Chen G, Chmura GL, Crooks S, Kairo JG, Liao B, Lin G, Troxler TG (2014) Chapter 4: Coastal wetlands. In: Hiraishi T, Klug T, Tanabe K, Srivastava N, Jamsranjav B, Fukuda M, Troxler TG (eds) 2013 supplement to the 2006 IPCC guidelines for national greenhouse gas inventories: wetlands. IPCC, Gland

Lau WWY (2013) Beyond carbon: conceptualizing payments for ecosystem services in blue forests on carbon and other marine and coastal ecosystem services. Ocean Coast Manag 83:5–14

Lewis RR III (2005) Ecological engineering for successful management and restoration of mangrove forests. Ecol Eng 24:403–418

Lovelock CE, Atwood T, Baldock J, Duarte CM, Hickey S, Lavery PS, Masque P, Macreadie PI, Ricart AM, Serrano O, Steven A (2017) Assessing the risk of carbon dioxide emissions from blue carbon ecosystems. Front Ecol Environ 15:257–265

Macreadie PI, Nielsen DA, Kelleway JJ, Atwood TB, Seymour JR, Petrou K, Connolly RM, Thomson ACG, Trevathan-Tackett SM, Ralph PJ (2017a) Can we manage coastal ecosystems to sequester more blue carbon? Front Ecol Environ 15:206–213

Macreadie PI, Olliver QR, Kelleway JJ, Serrano O, Carnell PE, Lewis CJE, Atwood TB, Sanderman J, Baldock J, Connolly RM, Duarte CM, Lavery PS, Steven A, Lovelock CE (2017b) Carbon sequestration by Australian tidal marshes. Sci Rep 7:44071. https://doi.org/10.1038/srep44071

Marbá N, Arias-Ortiz A, Masqué P, Kendrick GA, Mazarrasa I, Bastyan GR, Garcia-Orellana J, Duarte CM (2015) Impact of seagrass loss and subsequent revegetation on carbon sequestration and stocks. J Ecol 103:296–302

Mashayekhi Z, Danehkar A, Sharzehi GA, Majed V (2016) Coastal communities WTA compensation for conservation of mangrove forests: a choice experiment approach. Knowl Manage Aquat Ecosyst 417:20

Murray BC, Pendleton L, Jenkins WA, Sifleet S (2011) Green payments for blue carbon. Nicholas Institute for Environmental Policy Solutions, Durham

Osland MJ, Enwright NM, Day RH, Gabler CA, Stagg CL, Grace JB (2016) Beyond just sea-level rise: considering macroclimatic drivers within coastal wetland vulnerability assessments to climate change. Global Change Biol 22:1–11

Pendleton L, Donato DC, Murray BC, Crooks S, Jenkins WA, Sifleet S, Craft C, Fourqurean JW, Kauffman JB, Marbá N, Megonigal P, Pidgeon E, Herr D, Gordon D, Baldera A (2012) Estimating global "blue carbon" emissions from conversion and degradation of vegetated coastal ecosystems. PLoS One 7:e43542

Siikamäki J, Sanchirico JN, Jardine SL (2012) Global economic potential for reducing carbon dioxide emissions from mangrove loss. Proc Natl Acad Sci U S A 109:14369–14374

Spaulding MJ (2016) The new blue economy: the future of sustainability. J Ocean Coast Econ 2:8. https://doi.org/10.15351/2373-8456.1052

Thomas S (2014) Blue carbon: knowledge gaps, critical issues and novel approaches. Ecol Econ 107:22–38

Tramel S (2016) The road to Paris: climate change, carbon and the political dynamics of convergence. Globalizations 13:1–10. https://doi.org/10.1080/14747731.2016.1173376

Van Katwijk MM, Bos AR, DeJonge VN, Hanssen LSAM, Hermus DCR, de Jong DJ (2009) Guidelines for seagrass restoration: importance of habitat selection and donor population, spreading of risks, and ecosystem engineering effects. Mar Pollut Bull 58:179–188

Warner R, Kaidonis M, Dun O, Rogers K, Shi Y, Nguyen TTX, Woodroffe CD (2016) Opportunities and challenges for mangrove carbon sequestration in the Mekong River delta in Vietnam. Sustain Sci 11:661–677

Wilkinson C, Salvat B (2012) Coastal resource degradation in the tropics: does the tragedy of the commons apply for coral reefs, mangrove forests and seagrass beds. Mar Pollut Bull 64:1096–1105

Williams P, Faber P (2001) Salt marsh restoration experience in San Francisco Bay. J Coast Res 27:203–211

Wylie L, Sutton-Grier AE, Moore A (2016) Keys to successful blue carbon projects: lessons learned from global case studies. Mar Pol 65:76–84

Yee SM (2010) REDD and BLUE Carbon: carbon payments for mangrove conservation. Master's thesis, Center for Marine Biodiversity and Conservation, UC San Diego

Zarate-Barrera TG, Maldonado JH (2015) Valuing blue carbon: carbon sequestration benefits provided by the marine protected areas in Colombia. PLoS ONE 10:e0126627

Zedler JB (2000) Progress in wetland restoration ecology. Trend Ecol Evol 15:402–407

Chapter 7
Summary and Conclusions

Since the development of the concept of 'blue carbon' in 2009 to complete the global carbon accounting assessment begun by the IPCC, there has been rapid growth in the number of papers published dealing with the science and management of blue carbon. The focus of blue carbon is the coastal zone, namely, salt marshes, mangrove forests, seagrass meadows and possibly kelp forests. The science and management of these coastal ecosystems are being considered against a background of moderate to rapid deterioration and destruction of these highly valued and heavily used ecosystems.

Salt marshes are currently being destroyed at a rate of about 1–2% of area per year, and the current area is currently estimated at 41,657 km^2. Salt marshes are responsible for a variety of ecosystem services, such as commercial and recreational fishing and protection and mitigation against storm damage.

Sediment accretion in salt marshes has been commonly measured; the range of rates is 2–10 mm year^{-1} with a median of 5 mm year^{-1}. The single greatest driver of sediment, and thus carbon, accretion is the frequency of tidal inundation; the more frequently that tidal water containing silt and clay and associated carbon particles overlies the marsh surface, the greater the time available for such material to settle. Thus, accretion rates tend to decline with increasing tidal height and decreasing tidal inundation frequency.

Measurement of sediment accretion and carbon sequestration is not without its pitfalls as every method has its positive and negative aspects.

Salt marshes sequester carbon at a median rate of 184 g C_{org} m^{-2} year^{-1}, being a function not of latitude but of a number of interrelated factors such as marsh age, tidal inundation frequency and the magnitude of inputs from land and sea. Globally, the carbon sequestration rate for salt marshes is about 10 Tg C_{org} year^{-1}.

A median value of 282.2 Mg C_{org} ha^{-1} was calculated for carbon storage in salt marshes with a global figure of 1.2 Pg C_{org}, nearly all stored in soil. Carbon storage is greatest in mature marshes which underscores the fact that sufficient time is required for sediment and associated carbon to accumulate. Since the 1800s, approximately 25% of salt marshes have been destroyed, accounting for 1.1 Pg CO_2 equivalents (e) returned

© The Author(s) 2018
D. M. Alongi, *Blue Carbon*, SpringerBriefs in Climate Studies,
https://doi.org/10.1007/978-3-319-91698-9_7

to the atmosphere and/or the coastal ocean. Currently, assuming a global loss rate of 1–2% per year, carbon emissions from salt marsh losses are about 20–240 Tg CO_2 equivalents year^{-1}.

Like salt marshes, mangrove forests both actively and passively capture sediment and associated carbon particles from the overlying tidal water. Unlike salt marshes, mangroves can store significant quantities of carbon in above- and below-ground biomass. However, 75–95% of mangrove carbon is stored below-ground in dead roots. In the long term as in salt marshes, mangrove carbon is stored as peat. Peat formation is a function not only of rates and magnitudes of inputs but is also due to slow decomposition rates of mostly refractory matter, the tidal regime, micro- and macro-organism activities, tree species and litter composition, moisture and temperature. As a result of a combination of these factors, peat formation and accumulation occur in some salt marshes and mangrove forests but not in others.

Soil accretion rates in mangroves are similar to those measured in salt marshes, although most measurements are within a narrower range of 0–2 mm year^{-1} with a lower median figure of 3 mm year^{-1}. Carbon sequestration rates have a median value of 103 g C_{org} m^{-2} year^{-1} which equates to a global rate of 14.2 Tg C_{org} year^{-1}. As in salt marshes, the most mature mangroves sequester more carbon than younger, immature forests as there is no trend with latitude. Carbon sequestration is driven by factors other than forest age, such as tidal inundation frequency and inputs from land and sea.

Carbon stocks are greater in mangrove forests than in salt marshes or seagrass beds (Table 7.1), with a median value of 723.4 Mg C_{org} ha^{-1} for a global carbon stock of 10 Pg C_{org}. Ninety-two percent of this carbon is stored below-ground, but as for salt marshes and seagrass beds, this value is a likely underestimate as there is very probably much more carbon in soils deeper than 1 m. Potential carbon emission losses range annually from 0.27 to 0.59 Pg CO_2e year^{-1} globally, which adds 5–11% to the global deforestation rate.

Table 7.1 Estimates of carbon sequestered and stored in coastal ecosystems and their carbon emission and economic loss. e = equivalents

Ecosystem	Median carbon sequestration rate (g C_{org} m^{-2} d^{-1})	Median global carbon sequestration rate (Tg C_{org} year^{-1})	Median carbon stock (Mg C_{org} ha^{-1})	Median global carbon stock (Pg C_{org})	Carbon emissions (Pg CO_2e year^{-1})	Economic cost (billion US year^{-1})[a]
Salt marsh[b]	184	10.0	282.2	1.2	0.04–0.08	$0.52–$1.04
Mangrove[c]	103	14.2	723.4	10.0	0.27–0.59	$3.51–$7.67
Seagrass[d]	167	50.2–100.0	69.3	2.2–4.4	0.54–1.08	$7.02–$14.04

[a]Assumes a carbon market price of $13 US per Mg CO_2 equivalents (as of 1 June 2016)
[b]Assumes a global area of 41,657 km^2 and a global loss rate of 1–2% (see Chap. 2)
[c]Assumes a global area of 137,760 km^2 and a global loss rate of 1–2% (see Chap. 3)
[d]Assumes a range of global area of 300,000–600,000 km^2 and a global loss rate of 7% (see Chap. 4)

Seagrass meadows, like salt marshes and mangrove forests, are sites of passive and active accumulation of sediment and associated carbon. The median rate of carbon sequestration is 167.4 g C_{org} m^{-2} year^{-1} which is similar to the median value for salt marshes. There is a wide spread of data concerning soil carbon storage due to the fact that many studies, particularly early ones, did not core to the deepest possible sediment layers; most observations indicate carbon storage of <100 Mg C_{org} ha^{-1} with a median of 69.3 Mg C_{org} ha^{-1}. Nearly all carbon is stored in soil. Meadows of the Mediterranean seagrass *Posidonia oceanica* have by far the greatest carbon storage capacity with some *P. oceanica* sites having as much carbon as mature mangrove forests.

As global area is poorly known, global carbon storage in seagrass beds ranges from 2.1 to 4.2 Pg C_{org} and between 75.5 and 151 Tg in biomass. Assuming that the studies measuring soil carbon to a depth of at least 1 metre contain the complete inventory, carbon storage is then 5.8 to 9.8 Pg C_{org}. Due to their high rate of loss (7% per annum), 0.54 to 1.08 Pg C_{org} is lost annually to the atmosphere and the coastal ocean. This means that seagrass loss is equivalent to 9–25% of the annual deforestation rate.

Not enough data exists to determine whether or not kelp beds are a blue carbon store, but they have high potential due to their high rates of net primary productivity and detritus production. As kelps grow on rocky shores, it is unlikely that kelp beds themselves will be sites for storage but adjacent unconsolidated seabed areas may be. This idea may prove to be a fruitful avenue of research in the future.

The total amount of carbon sequestered and stored in coastal ecosystems and the economic cost of continuing losses of these habitats globally can be estimated by adding up the empirical data of salt marshes, mangrove forests and seagrass meadows from the earlier chapters.

Table 7.1 shows that salt marshes sequester more carbon, on average, than mangroves and seagrass beds although the range of values is not statistically significant. However, on a global basis, most organic carbon is sequestered by seagrass beds. The problem with estimating the impact of seagrass meadows in these calculations is our poor knowledge of their areal extent globally. Mangroves store more carbon than either salt marshes or seagrass meadows. Total coastal stocks of organic carbon range from 13.4 to 15.6 Pg C_{org} globally. This range is equivalent to only 1.6–1.8% of the world's forest carbon stocks (Pan et al. 2011). However, the range of carbon sequestration rates (74.4–124.2 Tg C_{org} year^{-1}) equates to about 3–5% of the global forest sink of 2.4 Pg year^{-1} (Pan et al. 2011). Depending on the global area of seagrass meadows, the estimated annual carbon emissions from habitat losses range from 0.85 to 1.75 Pg CO_2 equivalents. These estimates equate to 18 to 38% of annual CO_2 emissions from global deforestation (Hansen et al. 2013). Obviously, there are data gaps, and the level of uncertainty is high (Duarte 2017), but clearly coastal habitats require conservation and management urgently given the continuingly high rates of loss.

The financial costs of these losses are high, varying from $11.05 to $22.75 billion US annually, mostly from loss of seagrasses. These values compare favourably with the earlier estimates of Pendleton et al. (2012). While on an individual project basis, carbon trading may or may not be viable, clearly it is viable from a global perspective to conserve these habitats from further loss. More accurate knowledge of seagrass area is urgently needed to refine these estimates of carbon sequestration and storage and economic costs.

Salt marshes, mangrove forests and seagrass beds sequester and store more carbon on a per area basis than nearly all other ecosystems and clearly are prime sites to retain carbon as any losses back to the atmosphere or the coastal ocean are disproportionate to their small area compared with the world's forests. From an economic and global perspective, the main value in a blue carbon project lies in conserving what remains of these habitats. Blue carbon projects and their financing are still in their infancy, but given their importance, both ecologically and financially, blue carbon as a concept is likely to mature very quickly in the near future.

References

Duarte CM (2017) Reviews and syntheses: Hidden forests, the role of vegetated coastal habitats in the ocean carbon budget. Biogeosciences 14:301–310

Hansen MC, Potapov PV, Moore R, Hancher M, Turubanova TA, Thau D, Stehman SV, Goetz SJ, Loveland TR, Kommareddy A, Egorov A, Chini L, Justice CO, Townshend JRG (2013) High-resolution global maps of 21st-century forest cover change. Science 342:850–853

Pan Y, Birdsey RA, Fang J, Houghton R, Kauppi PE, Kurz WA, Phillips OL, Shvidenko A, Lewis SL, Canadell JG, Ciais P, Jackson RB, Pacala SW, McGuire AD, Piao S, Rautiainen A, Sitch S, Hayes D (2011) A large and persistent carbon sink in the world's forests. Science 333:988–993

Pendleton L, Donato DC, Murray BC, Crooks S, Jenkins WA, Sifleet S, Craft C, Fourqurean JW, Kauffman JB, Marbá N, Megonigal P, Pidgeon E, Herr D, Gordon D, Baldera A (2012) Estimating global "blue carbon" emissions from conversion and degradation of vegetated coastal ecosystems. PLoS One 7:e43542

Printed in the United States
By Bookmasters